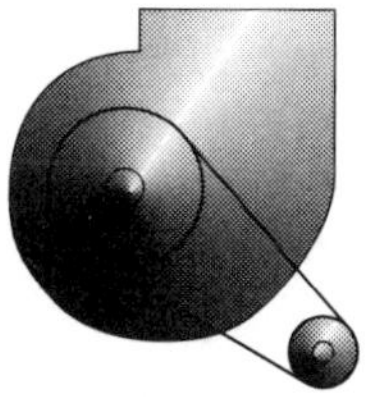

Fans and V-belt Drives

By

LEO A. MEYER

H. LYNN WRAY, P.E., Technical Advisor

INDOOR ENVIRONMENT TECHNICIANS LIBRARY

LAMA BOOKS

FOREWORD

You are probably working as a technician in one of the indoor environment fields. This means that you have probably worked with fans and V-belt drives. However, don't fall into the trap of thinking, "I know all this stuff."

Read each chapter. Then do the Review at the end of the chapter. In my experience, every time I studied material I "knew all about," I learned new ideas and corrected misunderstandings.

If you study each chapter carefully, you will gain new ideas. More important, you will give yourself a solid understanding of how to protect yourself on the job.

Indoor Environment Technician's Library

This book is part of the *Indoor Environment Technician's Library.* These are practical books that you can use for training or as reference. These books apply to all areas of the indoor environment industry, such as:

- Heating, ventilating, and air conditioning
- Energy management
- Indoor air quality
- Sheet metal work
- Building commissioning
- Testing, adjusting, and balancing

If you are a supervisor training others, you will find that the *Indoor Environment Technician's Library* can make it easier.

Leo A. Meyer

LAMA Books
Leo A. Meyer Associates Inc.
20956 Corsair Blvd.
Hayward CA 94545-1002
888-452-6244
510-785-1099

ISBN 0-88069-026-7

TABLE OF CONTENTS

1 FAN TERMS

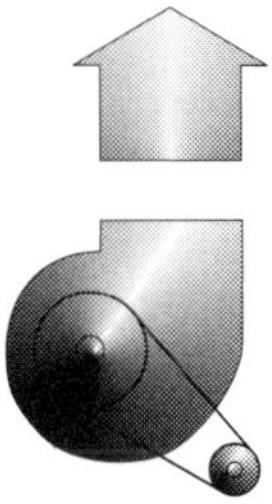

FANS

Many people think of fans as the propeller blade type of fan that is used to create a draft in a room. To the indoor environment technician, the term **fan** means all types of devices that move air. The fan is also called a **blower** or an **AMD** (air moving device). The fan most commonly used for HVAC work (heating, ventilating, and air conditioning) is the **centrifugal fan** (Fig. 1). Types of fans are described in Chapter 3.

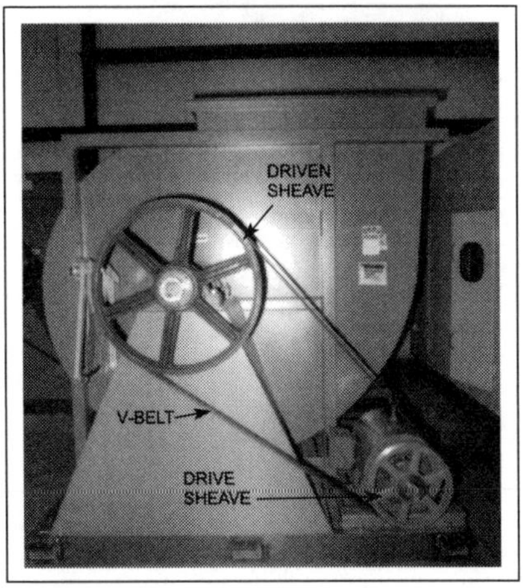

Fig. 1: Centrifugal fan

The terms used with fans vary in different areas. Look over the following terms to see if they are the terms used in your area:

- **Equipment room**—A room where the central HVAC equipment is installed (piping, pumps, fan plenum, etc.)
- **Fan plenum**—A large sheet metal enclosure for the fan and other items in the equipment room (Fig. 2). It is airtight in order to direct the air path through the filters, coils, and fan. Often the fan plenum is large enough for a person to enter and to stand upright in order to service the fan, coils, and filters.

Fig. 2: Fan plenum

- **V-belt drive** (Fig. 1): The system that transmits power from the motor to the fan. It has three parts:
  - **Drive sheave**: The sheave on the motor.
  - **Driven sheave**: The sheave on the fan shaft.
  - **V-belt**: The belt that fits on the sheaves and transmits power from the motor to the fan. A small fan may have only one V-belt. Large fans have multiple V-belts.
- **Sheaves** are sometimes called **drives** or **pulleys**. In the HVAC industry, sheave is pronounced *shiv* (rhymes with *give*).

AIR PRESSURE

A fan is similar to a water pump. Both create a pressure difference in a system in order to make a fluid flow. Both water and air are fluids, and they act much the same way. A fan creates a higher pressure at the outlet than at the inlet. This pressure difference causes the air to flow. The key to understanding fans and their performance is understanding airflow and air pressure produced by the fan in the ductwork connected to it.

We can't see air, but we know it has **weight** and **pressure.**

Even if air is not moving, it has **weight** and presses against any object. For example, the standard barometric pressure of air at sea level is 29.92 inches of mercury, or 14.7 psi (pounds per square inch). This pressure of the air which is all around us is called **atmospheric pressure**.

We feel the **pressure** of air when the wind blows. A draft is a small amount of pressure, but a storm may create a very high pressure wind. Air tends to move from a higher pressure to a lower pressure. When the air pressure in one geographic area becomes lighter (perhaps because the sun has heated it), air will rush in. The rush of air is wind.

A fan moves air by creating a relatively high pressure air at the fan discharge. The higher pressure air moves to fill a space that has a lower pressure. The air pressure created by a fan is relative to atmospheric pressure:

- **Positive fan pressure** at the fan discharge is higher than atmospheric pressure. This positive pressure moves air through the ductwork.
- **Negative fan pressure** at the fan inlet is lower than atmospheric pressure. A negative pressure is created at the fan inlet when air is pulled into the fan.

The air moving in ductwork has both static pressure (SP) and velocity pressure (VP). The sum of these two pressures is the total pressure (TP).

Static Pressure (SP)

It is a law of physics that whenever a contained fluid (air is a fluid) is under pressure, it exerts the pressure equally in all directions. This is called **static pressure** (**SP**) because it can exist without air movement. A balloon stays inflated because the static pressure is pushing on the inside of the balloon in all directions.

When a fan is moving air through a duct, static pressure exerts pressure on the inner sides of the duct. If the fan is turned off and no air is moving in the duct, the static pressure inside the duct is the same as the atmospheric pressure (14.7 psi at sea level).

Velocity Pressure (VP)

Velocity pressure (**VP**) is the pressure that results from the air being in motion. Velocity is the rate of airflow (the speed). The greater the velocity (speed) of the air, the greater the velocity pressure.

Total Pressure (TP)

Total pressure (**TP**) is the total air pressure that is present in a duct system. It is the sum of the velocity pressure (VP) and the static pressure (SP). The relationship between these can be expressed in any one of these equations:

TP = SP + VP

SP = TP – VP

VP = TP – SP

These equations are useful when working with fan and airflow data. If you wish to learn more about static pressure and velocity pressure, see the book *Airflow in Ducts* in this *Indoor Environment Technician's Library.*

Measuring Air Pressure

Air static pressure in a duct is measured by how high the air pressure can raise a column of water in a U-tube when one side is open to the atmosphere:

- ❑ Figure 3 shows a U-tube that contains water. The water level is the same in both tubes because atmospheric pressure affects them both.
- ❑ Figure 4 shows a flexible hose connected to one side of a U-tube inserted in a duct. When the pressure of the air in the duct

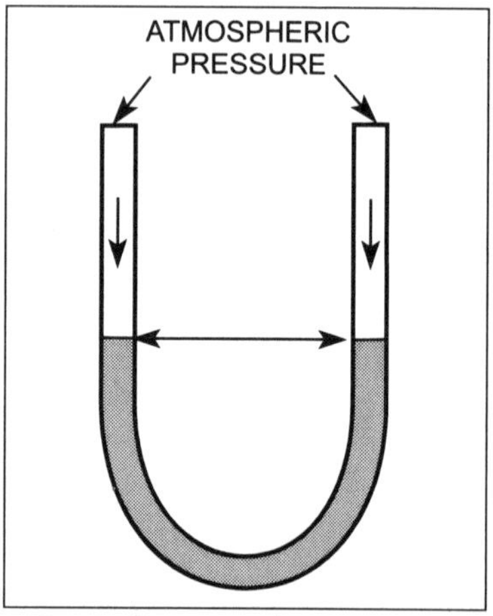

Fig. 3: Because the air pressure in both tubes is equal, the water levels are equal

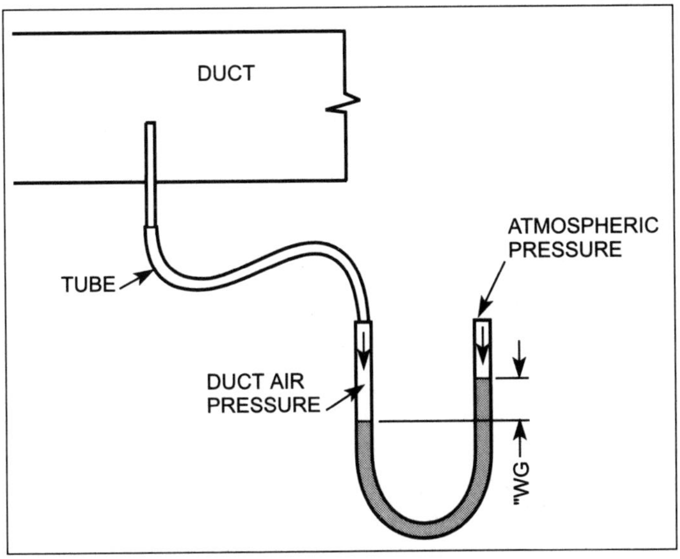

Fig. 4: Unequal liquid levels when the air pressures are different

pushes against the water on one side of the U-tube, the water rises on the other side.

The difference in inches between the two water levels is called **inches water gage**, or **"wg**. It may also be labeled **inches of water**, **"H_2O**, or **inches water column** (**"WC**). Some U-tubes are filled with a special oil instead of water, but the scale on the tube is changed so that the amount is still read in "wg (inches water gage).

Devices that measure the air pressure in a duct are called **manometers**. They do not always have the U-tube, but the measurement of air pressure is still given as "wg.

Because both air and water are fluids, the pressure of air moving through duct can be compared to the pressure of water moving through pipe:

- ❑ Pump pressure is measured in **feet of water** or in **pounds per square inch** (**psi**). The pressure in a domestic water line is usually about 30 psi (pounds per square inch).
- ❑ Fan pressure is measured in **"wg** instead of psi because the relatively small pressures in a duct system are too small to be measured accurately in psi. For example, 1"wg is a common pressure for fans

and ductwork. A fan pressure of 1"wg is equal to only four hundredths of one psi (0.04 psi).

TERMS USED WITH FANS

Before studying fans, you need to understand the terms used to describe air movement and fan performance.

CFM (cubic feet per minute)

The airflow rate is the volume of air that the fan delivers in a given time. It is usually measured in **cubic feet per minute** (**CFM**). Visualize one cubic foot as a block of air measuring 1 foot x 1 foot x 1 foot.

FPM (feet per minute)

The **velocity** is the speed of air flowing through a duct or out the discharge of a fan. It is measured in **feet per minute** (**FPM**).

OV (outlet velocity)

Outlet velocity (**OV**) is the velocity of the air at the fan outlet. Fan tables usually list the fan OV at a given airflow rate (CFM).

RPM (revolutions per minute)

The rotational speed of the fan wheel is measured in **revolutions per minute** (**RPM**). It is usually measured at the center of the fan shaft. However, the RPM applies to any fixed point on the fan wheel.

Tip Speed

Tip speed is the speed that a point on the outer edge of a fan wheel is traveling. Excessive tip speed may damage the wheel or bearings. Tip speed is measured in FPM (feet per minute). RPM is usually used for fan data to designate rotational speed, but tip speed is sometimes referred to. Manufacturers use it in design of fans. Tip speed is not often used in the field, but you should know what it means.

Tip speed can be calculated by multiplying the fan RPM times the outer circumference of a fan wheel:

Tip speed = Fan RPM x Circumference (ft) of fan wheel

For example, if a fan wheel is 11½" in diameter, its circumference is approximately 3 feet (because Circumference = π x Diameter). This means that in one revolution of that fan shaft, a point on the outer edge of a blade is traveling 3 feet. At 1000 RPM, the tip speed for this fan wheel is 3000 FPM:

Tip speed = 1000 RPM x 3 ft. circumference

Tip speed = 3000 FPM (feet per minute)

Horsepower (Hp)

Horsepower (**Hp**) is a unit of measurement that indicates the work capacity of a motor. Electric motors for fans are rated in horsepower. Motors for small fans may be **fractional** (less than 1 Hp). Small commercial fan motors may be in the range of 3 to 10 Hp. Very large fan motors may be 75 Hp, or even more for very large installations.

Brake Horsepower

Brake horsepower (**Bhp**) is the load imposed on a motor by the fan. That is, it is the power required at the fan shaft, plus the V-belt power loss. If a fan motor is 2 Hp, but the power required by the fan is 2.5 Bhp, the motor is not large enough and it will be overloaded.

Fan manufacturers' data gives the Bhp required at a given CFM and static pressure for a particular fan. For example, the set of conditions for a fan is:

CFM – 27,072

SP – 2.0 "wg

RPM – 485

Bhp – 11.7

Under these operating conditions, a 15 Hp motor is needed. The motor horsepower must be large enough to overcome the fan Bhp, plus have extra horsepower for V-belt drive loss and sufficient torque to overcome the inertia of the fan during start-up.

Torque

A force transferred by a twisting or rotating motion.

Inertia

Tendency to remain at rest or to remain in motion.

Efficiency

A fan is never 100% efficient. Not all of the motion of a fan produces the work of moving air. For example, some of the work goes into overcoming the friction of the moving parts. **Fan efficiency** is the relationship between useful work produced by the fan and total work input by the fan motor:

$$\text{Fan efficiency} = \frac{\text{Useful work output}}{\text{Total work input}}$$

There is more information on efficiency in Chapter 4, Fan Curves and Fan Tables.

System Effect

System effect is the negative effect on fan performance

when a fan is attached to a duct system. The data given by the manufacturer for a particular fan (such as the CFM and Bhp at a given RPM) is established by laboratory testing. Once the fan is attached to the duct system, the configuration of the fan installation and duct connections affects the fan performance. System effect cannot be measured in the field.

The increased resistance due to system effect changes the fan performance. The degree of change depends upon the design and size of the duct system. Therefore actual fan performance can only be determined by measurements in the field. Chapter 5 discusses system effect in detail.

REVIEW

If you can match the following items without referring to the text, you have learned the contents of this chapter. Try to match every item before you check the answers in the back of the book.

Matching

____ 1. SP
____ 2. VP
____ 3. TP
____ 4. "wg
____ 5. CFM
____ 6. FPM
____ 7. OV
____ 8. RPM
____ 9. Tip speed
____10. Hp
____11. Bhp
____12. Fan efficiency
____13. System effect

A. FPM at fan outlet
B. SP + VP
C. Volume of air delivered by a fan
D. Speed at rim of fan wheel
E. Pressure exerted in all directions by air in a duct
F. Motor or fan speed
G. A measure of SP
H. Pressure exerted by air speed
I. Air velocity
J. Work capacity of a motor
K. Results from connecting duct to a fan
L. Work required by a fan
M. Work output divided by work input

2 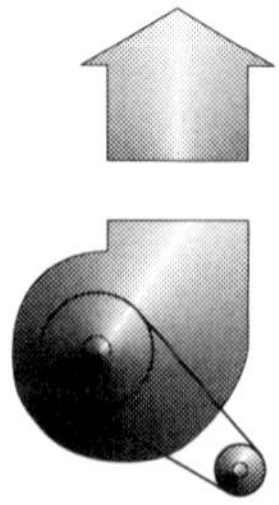

FAN SAFETY

MOVING PARTS

The moving parts of a fan can cause many accidents. Follow these rules for working on fans, changing V-belt sheaves or bolts, or working on any moving equipment.

Lock Out/Tag Out

The only sure way to prevent a fan from starting unexpectedly is to lock out the fan motor. This means turning off the main disconnect switch and putting a padlock on it so no one but you can turn it on. Use this procedure:

- ❑ Disconnect the main switch to the motor and put your padlock on it. You should be the only one with a key to the padlock. Attach a note or a tag to the lock stating who is working on the machinery. Put the date and time on the tag, so no one can think it is an old tag.
- ❑ Turn off the manual ON-OFF switch that should be within access of the fan motor.
- ❑ Test the power to the fan motor to make sure the line is dead. A switch can be wired incorrectly and still send power to machinery even when it is in the OFF position. If the line is not dead, the fan may start automatically.

If the motor is controlled by a variable frequency drive (VFD), do not throw the disconnect when the motor is running. Stop the fan first. Throwing the disconnect while the fan is running can damage the VFD. Variable frequency drives are discussed in Chapter 7.

Block Out

A worker who was changing the V-belts on a return-exhaust fan de-energized the fan motor before starting work. However, he did not block the blades of the fan, so the wheels could still move. The suction in the ductwork created by the supply fan caused the fan wheel to turn and his hand was caught in the V-belt drive.

This example shows the importance of blocking a fan wheel or any mechanical part that could start moving even with the power off. A good way to prevent fan rotation is to install a wood block at the fan wheel in a way that will not do any damage. **BE SURE TO REMOVE THE BLOCK WHEN YOU ARE THROUGH**.

Working with Moving Parts

Never try to stop a V-belt drive with your hands, no matter how slowly it is moving. When the power to a fan is turned off, the fan coasts to a stop. Let the drive stop **completely** before you touch it. The V-belt and sheaves may be moving so slowly that they appear easy to stop. But no matter how slowly the belt and sheaves are moving, you cannot stop them by hand. The combined momentum of the fan, motor, and drives is too strong. Your hands can be torn or crushed by being pulled between the V-belt and the sheave. Some sheaves may have protruding set screws. These will tear your hand even if there is only a slight rotation.

Never—under any circumstances—put your fingers under V-belts. When working on a V-belt drive, even with the equipment locked out, it is possible to move the fan sheave slightly and have your fingers caught between the belts and the sheave. If the equipment somehow starts and your fingers are under the belts, they will be badly injured.

Belt Guards

A fan must have a **belt guard** (Fig. 1) over the V-belt drive to prevent hands or clothes from being caught. You often have to remove the guard to do your work. Always replace the guard before starting the fan.

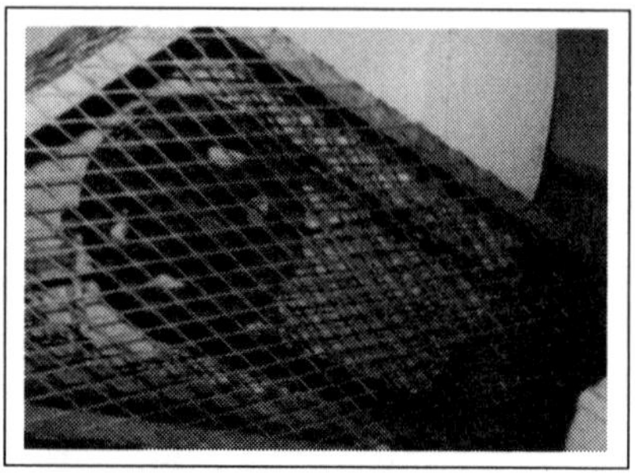

Fig. 1: A belt guard must be in place over the V-belt drive

If the guard must remain off temporarily while the machinery is running, put a barrier (such as a ladder) in front of it. This warns anyone that there is danger. The barrier also reminds you of the danger, because it is easy to forget that the guard is off. The barrier is especially important in a fan plenum. This is often a dimly lit place where it is not easy to see the exposed machinery.

IN THE FAN PLENUM

Fan plenums are frequently dimly lit and have very little extra room. In addition, they have moving parts such as V-belt drives. They require special safety precautions:

- ❑ Tell someone that you are entering a fan plenum. If you get trapped there, you need someone to know where you are.
- ❑ Do not wear loose clothing or long sleeves that can catch on moving parts.
- ❑ If you have long hair, keep it tied up or under a cap.
- ❑ Carry a flashlight. You will need it in dimly lit areas.

- ❑ Often there is little head room in a fan plenum. When you enter, use a flashlight to look for all the possible places where you could hit your head. Electric lights in fan plenums may have protective and waterproof covers which are often very low.
- ❑ Carry a cell phone, especially if you are working alone, so that you can call for help if you are injured or trapped in a fan plenum.

Always open the fan plenum doors carefully. Excess pressure can blow the doors open violently on the fan discharge side.

You can be trapped in a fan inlet plenum. If there is no fan shut-off switch in the plenum and the outside air and return air dampers accidentally close, the fan will create a negative pressure in the plenum. This pressure can easily be strong enough to prevent you from pushing open the plenum door.

Before working in a fan plenum, remove all debris and stored material. The plenum is sometimes a favorite place to store material. Anything stored in the plenum can contaminate the conditioned air. It also increases your chance of tripping.

Make sure you have all the tools you need before you enter the plenum. Once in an awkward position, it is so much trouble to get out and get the proper tool that you will be tempted to do foolish things such as using a knife for a screwdriver or using an adjustable wrench for a hammer. These lead to injuries, damaged tools, and a poorly done job.

Electrical Safety in the Fan Plenum

Follow all electrical safety rules carefully. Refer to the IET book *Safety for the Indoor Environment Technician* for electrical safety instruction.

Electrical safety is especially important when you are working in a fan plenum:

- ❑ Because you are working in close quarters, you can easily make an electrical ground against metal. This increases the hazard of electrical shock.
- ❑ The fan plenum must have a disconnect switch within sight of the fan motor.
- ❑ Avoid using extension cords in the fan plenum. Use battery-powered tools instead. If you must use cords, be very careful that they remain clear of any moving machinery.

REVIEW

If you can answer the following questions without referring to the text, you have learned the contents of this chapter. Try to answer every question before you check the answers in the back of the book.

1. Under what conditions can you stop a V-belt with your hands?

___

2. If a fan motor is running without the guard in place and the guard cannot be replaced immediately, what safety measure should you take?

___

3. What is the first action to take before starting work on a V-belt drive system?

___

4. Why should you never put your fingers under V-belts?

___

5. If you must work by yourself in a fan plenum, what safety measures should you take?

___

___

___

3 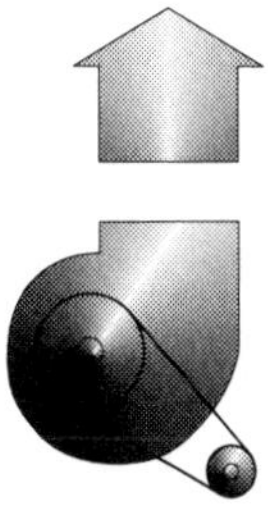

TYPES OF FANS

This chapter will describe two general types of fans:

- ❑ Centrifugal fans
- ❑ Axial fans

For any fan, four values can be used to describe its operating characteristics:

- ❑ Air quantity delivered (CFM)
- ❑ Static pressure (SP)
- ❑ Revolutions per minute (RPM)
- ❑ Brake horsepower (Bhp)

These terms were discussed in Chapter 1. Usually the designer knows the CFM and the SP needed for a particular duct system. The fan RPM and Bhp will vary according to the fan size and type as well as the required CFM and SP.

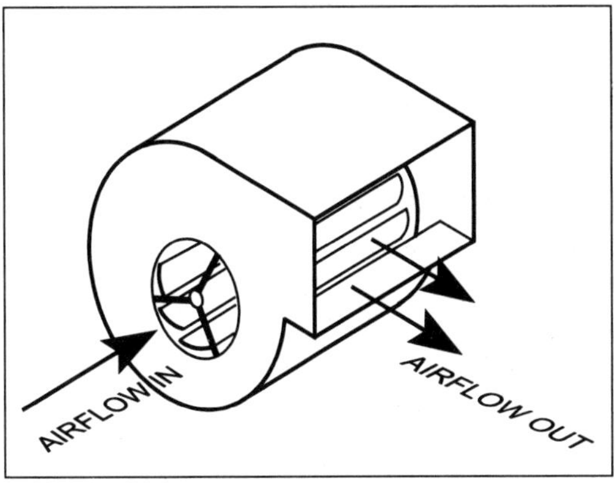

Fig. 1: In a centrifugal fan, air travels out at 90° to the centerline

CENTRIFUGAL FANS

Centrifugal fans discharge air by centrifugal force. At the outlet, the air travels out at 90° to the fan shaft centerline (Fig. 1). Centrifugal fans are also called **blowers** or **squirrel**

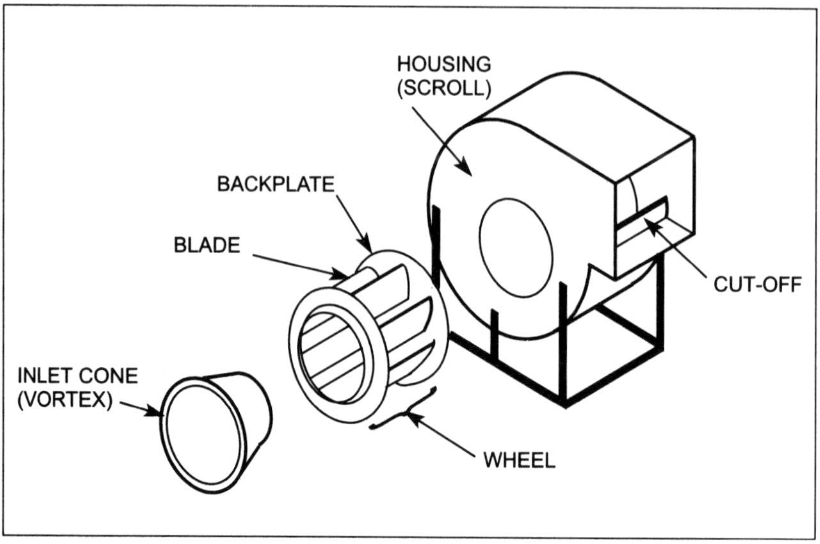

Fig. 2: The parts of a fan

cage fans. The fan components have these functions (Fig. 2):

- The **housing** (also called the **scroll**) is the fan casing. The scroll is engineered for the best performance for the particular fan model.
- The **cut-off** is a piece of metal that prevents the air discharged from the wheel from being drawn back into the scroll. This would reduce fan performance.
- The **wheel** includes the blades, the backplate, and the framework that holds them in place. The design of the blades depends on the purpose of the fan.
- The **inlet cone** (also called the **vortex**) is designed to efficiently draw the intake air into the wheel. Anything that obstructs the air entering the fan inlet decreases the fan performance.

The performance of a centrifugal fan depends upon a number of design factors:

- Design of the scroll

- Design of the blades
- Shape of the outlet
- Relation of inlet diameter to outlet dimensions
- Location and size of the cut-off

TYPES OF BLADES

Different types of centrifugal fans are designed for various uses. The design of the blades often determines how a fan is best applied.

Forward-Curved-Blade Fan

The **forward-curved-blade fan** (**FC**) (Fig. 3) is commonly used in small air conditioning systems such as residential systems. The curved blades can load up with dirt and cause the impeller to become unbalanced. Because of this, the forward-curved-blade fan should not be used for material or fume handling duct systems. It has these characteristics:

- There are usually 60 blades.
- If properly sized, it is the smallest fan for a given duty.
- Best for duct systems with low or medium static pressures.
- Best applications are for less than 20,000 CFM and for static pressures less than 1.50" wg.
- Relatively low space requirements.
- Low RPM and tip speed.
- Longer bearing life.

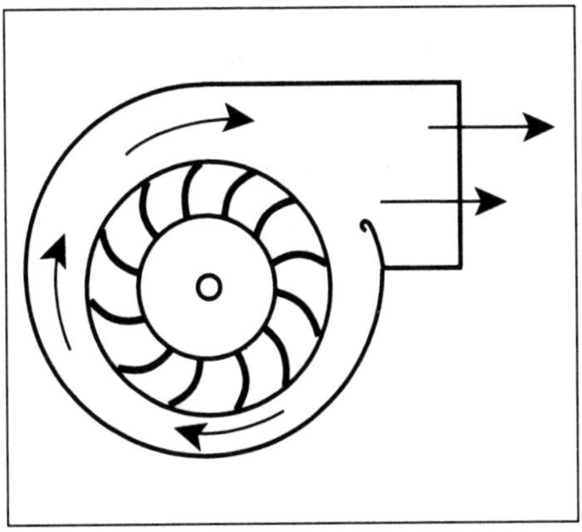

Fig. 3: Forward-curved-blade fan

- ❑ Quiet.
- ❑ Unstable performance if oversized.
- ❑ V-belt drive is usually more costly because a larger fan sheave is needed.
- ❑ Not good for moving dust or fumes. These can load up the blades, resulting in an unbalanced impeller.
- ❑ Bhp increases with increasing air quantity that is caused by reduction of system resistance. This can overload the motor.

Backward-Inclined-Blade and Backward-Curved-Blade Fan

The **backward-inclined-blade fan** (**BI**) and the **backward-curved-blade fan** (**BC**) (Fig. 4) have blades that slant away from the direction of rotation. The backward-inclined blade is straight instead of curved.

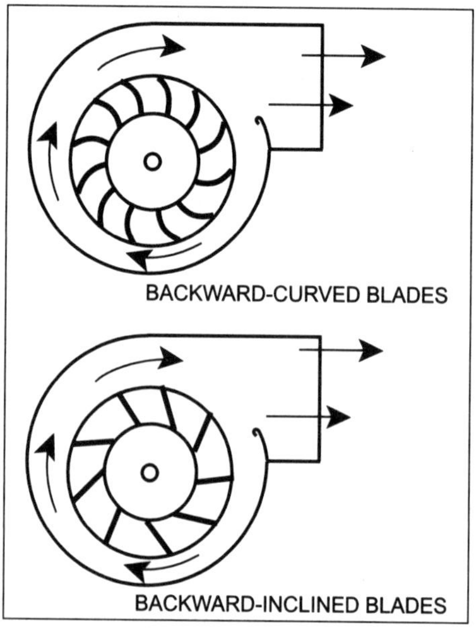

Fig. 4: Backward-curved and backward-inclined-blade fans

These blades are called "non-overloading" because if the motor is selected properly, it will not exceed its rated amps (become overloaded) regardless of the amount of airflow The fans have these characteristics:

- ❑ Generally used for medium to high static pressure range.
- ❑ More efficient than the forward-curved-blade fan.
- ❑ More stable than the forward-curved-blade fan.
- ❑ Relatively quiet at high speeds.

- ❑ Generally require larger shaft and bearing sizes than the forward-curved-blade fan.
- ❑ RPM is about twice as much as for forward-curved-blade fans for the same air quantity.
- ❑ The Bhp requirement is less than that for the forward-curved-blade fan for a given application.
- ❑ The Bhp increases with increasing CFM up to about peak efficiency and then decreases.

Airfoil-Blade Fan

The **airfoil-blade fan** (Fig. 5) is a variation of the backward-inclined-blade fan. The airfoil blade permits smoother airflow through the fan. It is the most efficient of the centrifugal fans. However, it has a slightly higher cost. Therefore it is generally used for high static pressure and high airflow rates, where the power savings will outweigh the higher cost of the fan.

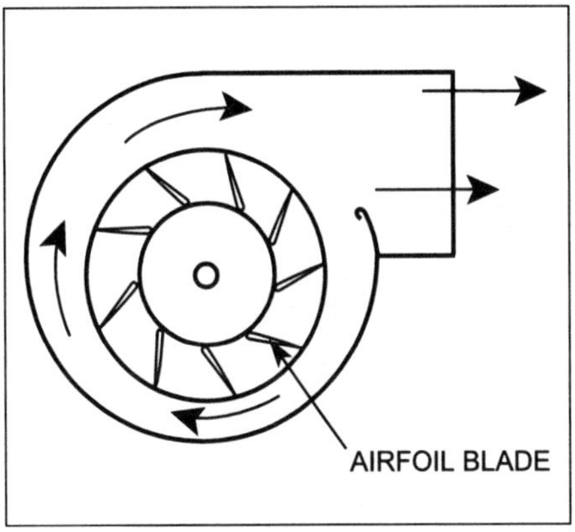

Fig. 5: Airfoil-blade fan

The characteristics of the airfoil-blade fan are similar to those of the backward-inclined-blade fan, except that it has a smaller range of high efficiency operation.

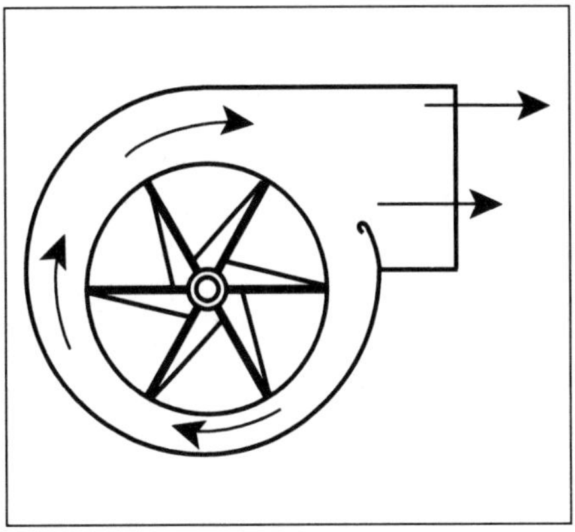

Fig. 6: Radial-blade fan

Radial-Blade Fan

The **radial-blade fan** (Fig. 6) is also called a **paddle-wheel fan**. It is commonly used for material handling systems such as sawdust collection or any application where a heavy dust load passes through the blades. The blades are less subject to material loading. (Material loading results in wheel imbalance.) The radial-blade fan has these characteristics:

- Less loading of materials on blades.
- Generally medium speed fans.
- Medium sound generation.
- Moves small volumes at high pressures.
- When CFM increases, Bhp increases and SP decreases.
- Least efficient of all centrifugal fans.
- Usually of heavier construction than other centrifugal fans.

Tubular Centrifugal Fan

The **tubular centrifugal fan** (Fig. 7) does not have the usual scroll casing. Instead, the fan wheel is installed in a tubular housing. The discharge air leaves the fan wheel at 90° to the fan shaft. However, guide vanes turn the air so that it is discharged from the outlet parallel to the fan shaft. The blades are backward-curved or airfoil design. The fan has these characteristics:

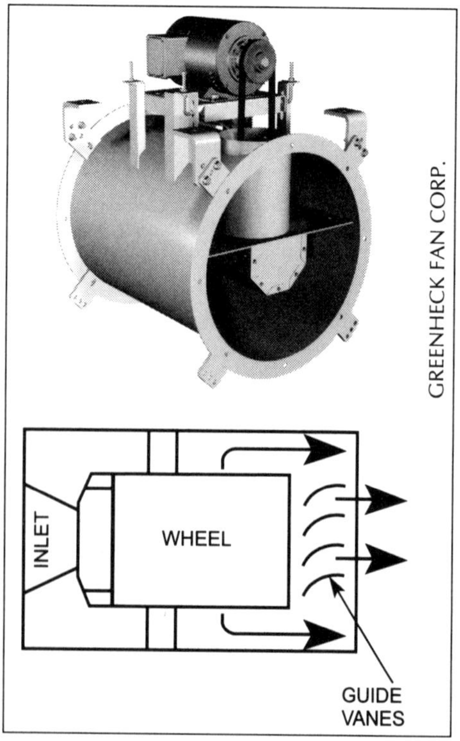

Fig. 7: Tubular centrifugal fan

- Usual use is roof exhaust fans for low pressure exhaust systems, such as kitchens, factories, and warehouses.

- Requires less space for installation.
- Produces a high volume of air against a low static pressure.

Plug Fan

The **plug fan** (Fig. 8) is a centrifugal fan that does not have a housing scroll. It is installed in the fan plenum of an air handling unit and pressurizes the entire space. This has two advantages:

Fig. 8: Plug fan

- The fan does not require a duct from the fan to the plenum wall.
- The discharge duct can be attached to any part of the plenum wall. It does not have to be in line with the fan.

The plug fan has these characteristics:

- Less efficient than other centrifugal fans.
- Used to save space.

DESIGN DESIGNATIONS FOR CENTRIFUGAL FANS

Width and Inlet Designations

Centrifugal fans that have only one wheel are called **single width fans** (**SW**). Single width fans have an inlet on one side only, so they are **single inlet** (**SI**). These fans are designated **SWSI fans** (Fig. 9).

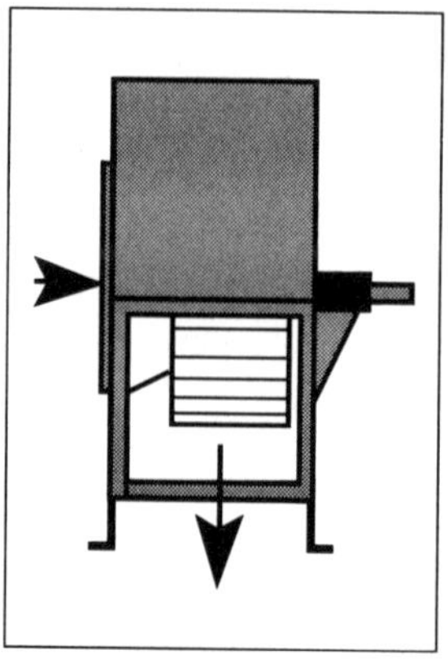

Fig. 9: Single width, single inlet fan (SWSI)

Centrifugal fans that have two wheels mounted side by side on the same shaft are called **double width fans** (**DW**). Double width fans have an inlet on both sides, so they are **double inlet** (**DI**). These fans are designated **DWDI fans** (Fig. 10).

Impeller Rotation

Fan rotation is designated as **CW** (**clockwise**) or **CCW** (**counterclockwise**). This always assumes that the viewer is facing the drive side of the fan (the side that has the V-belt drive).

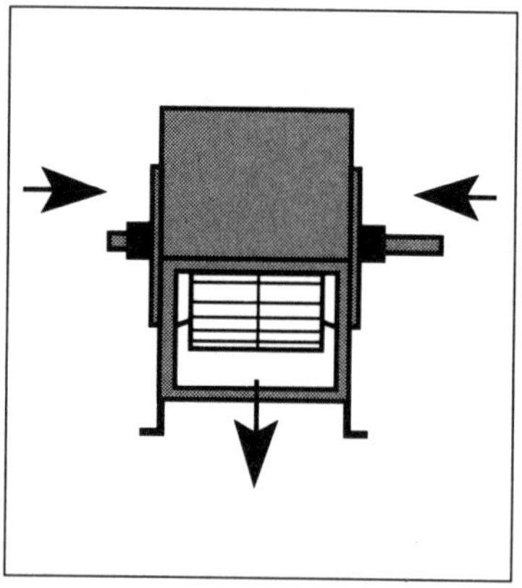

Fig. 10: Double width, double inlet fan (DWDI)

Direction of Airflow

The location of the fan discharge outlet can be changed on smaller centrifugal fans by removing the bolts from the mounts and rotating the housing. Larger fans have a fixed arrangement. As with fan rotation, the position designation assumes the viewer is facing the drive side. Figure 11 shows the various positions **for clockwise fans** (**CW**). The position is given by letters:

- TH – Top horizontal
- BH –Bottom horizontal
- UB – Up blast
- DB – Down blast
- TAU – Top angular up
- TAD – Top angular down
- BAU – Bottom angular up
- BAD – Bottom angular down

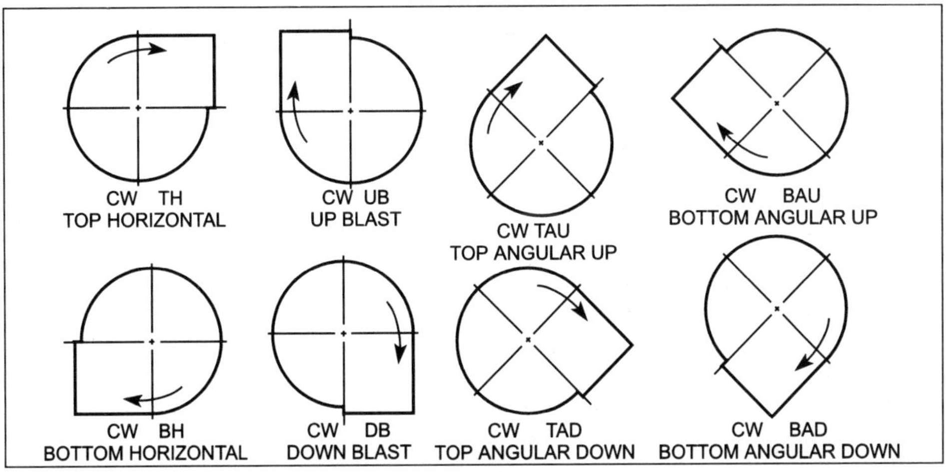

Fig. 11: Position designation for clockwise (CW) fans viewed from drive side.

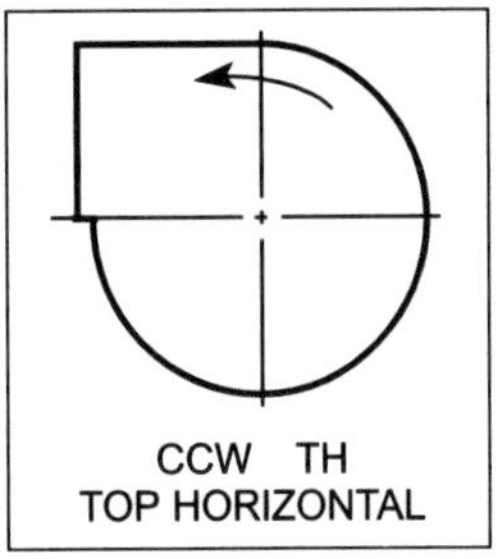

Fig. 12: Counterclockwise fan (CCW), top horizontal (TH)

Fans that rotate counterclockwise are designated CCW, with the same letters for the outlet position. Figure 12 shows a fan with a CCW TH designation.

Classes of Construction

Centrifugal fans are built to conform to standards set by **AMCA** (**Air Movement and Control Association**). AMCA has established fan classes I, II, III, and IV. The classes are based on a combination of **fan outlet velocity** in FPM and **static pressure** in "wg. The fan class specifies how heavy the fan construction must be in terms of such things as the gage of the housing and wheel, and the size of the shaft and bearings. Most fans in HVAC applications are either Class I or Class II.

In balancing a duct system, the fan is speeded up in order to increase the volume of air delivered. If the fan RPM is increased, it is possible to go beyond the limits for that class of construction. In such a case, the fan wheel may break apart, bearings can fail, or the fan shaft can whip, resulting in excessive vibration.

Fan tables published by the manufacturer usually indicate the limits of the CFM, static pressure, and RPM for the class of construction. Most fan tables are colored or shaded to distinguish between the different classes. Fan tables are covered in Chapter 4.

AXIAL FANS

Axial fans have the airflow parallel to the fan shaft, like common house fans. The three general types of axial fans are **propeller fans**, **tubeaxial fans**, and **vaneaxial fans**.

Propeller Fan

The **propeller fan** includes portable house fans. These fans have two or more blades. Four-blade and six-blade fans are probably the most common. All sizes are available from 6" diameter or less, up to about 7 feet in diameter. Smaller sizes are **direct drive** (the fan is attached directly to the motor shaft). Larger sizes are **belt driven**.

Propeller fans have some sort of **shroud** (Fig. 13) to guide the air through the fan blades. Most shrouds are bell-shaped at the inlet side to guide the air into the fan. The small diameter of the shroud is as close as is practical to the blade tips so that all the air is delivered through the fan blades. The design of the shroud substantially affects the fan performance.

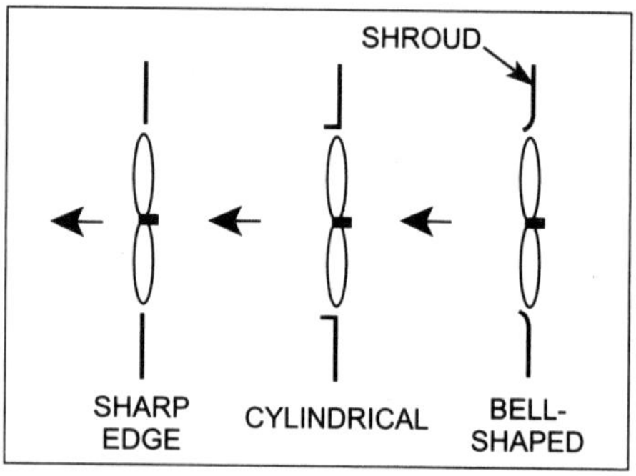

Fig. 13: Fan shrouds for propeller fan

Propeller fans are used in applications for high volume air and low SP (static pressure). As resistance against the fan increases, airflow is reduced significantly and the Bhp increases. For this reason, propeller fans are seldom connected to duct work. Typical applications are ceiling fans, attic fans, roof exhaust fans, and fans for moving air within a space.

Tubeaxial Fan

The **tubeaxial fan** (Fig. 14) is similar to a propeller fan, but it has a tube-shaped housing. It has 4 to 8 blades that are either airfoil or flat design. Common sizes are from 18" to 60" diameter.

Fig. 14: Tubeaxial fans

The tube increases the fan performance. The tube is designed to just clear the blade tips. The airflow at the fan outlet is spiral, but the flow straightens out a short distance from the outlet.

Tubeaxial fans can be **direct drive** or **belt driven**. The motor in a direct drive fan is in the air stream, so it is used for relatively clean air that is free of corrosion. Belt driven tubeaxial fans are typically used for paint spray booth exhaust and other fume exhaust.

Vaneaxial Fan

The **vaneaxial fan** (Fig. 15) is similar to a tubeaxial fan except that guide vanes are added before the fan. These vanes straighten the airflow and improve the fan efficiency and static pressure capabilities. The vanes make this the most efficient axial fan. It is often used in duct systems in place of the centrifugal fan, especially where space is limited. Vaneaxial fans have a very high noise level. They are often provided with some type of sound suppressor.

Fig. 15: Vaneaxial fan

The angle of the vanes is called **pitch**. The blades of a vaneaxial fan can be **fixed pitch** or **variable pitch** (also called **controllable pitch**). Variable pitch fans are especially useful in VAV (variable air volume) systems. Sensors for airflow, pressures, temperature, and other air conditions can be installed in the ductwork to control the pitch of the blades.

REVIEW

If you can answer the following questions without referring to the text, you have learned the contents of this chapter. Try to answer every question before you check the answers in the back of the book.

Matching

____ 1. Housing
____ 2. Cut-off
____ 3. Inlet cone
____ 4. Vortex
____ 5. Forward-curved-blade fan
____ 6. Backward-curved-blade fan
____ 7. Airfoil-blade fan
____ 8. Radial-blade fan
____ 9. Tubular centrifugal fan
____10. Plug fan

A. No housing
B. Fan casing
C. Commonly used for small centrifugal fans
D. Generally used for high static pressure and high airflow rates
E. Designed for smooth airflow into the fan
F. Used for material handling systems
G. Called "non-overloading" blades
H. Prevents air from being drawn back into the casing
I. Discharges air parallel to the wheel shaft

Identify these abbreviations:

11. SWSI

____

12. DWDI

____

Give the name and the letter designation for the clockwise rotation fans shown in the following drawings.

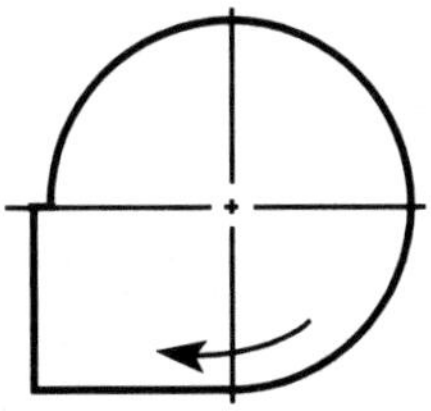

13. ____________________

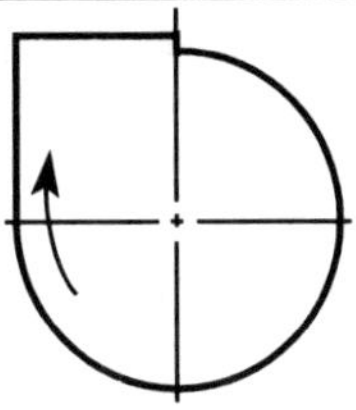

14. ____________________

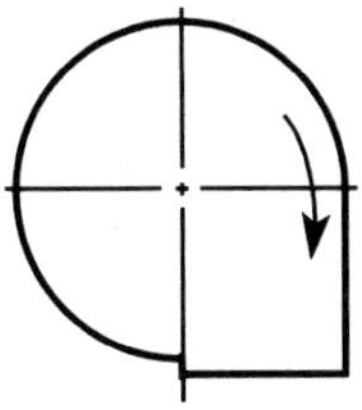

15. ____________________

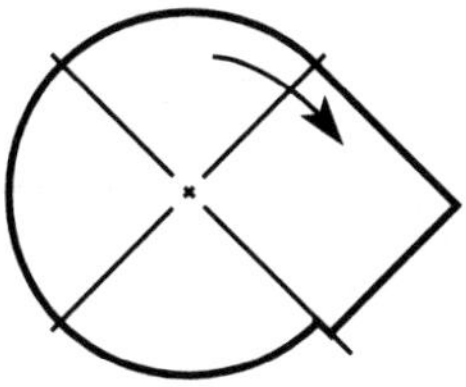

16 . ____________________

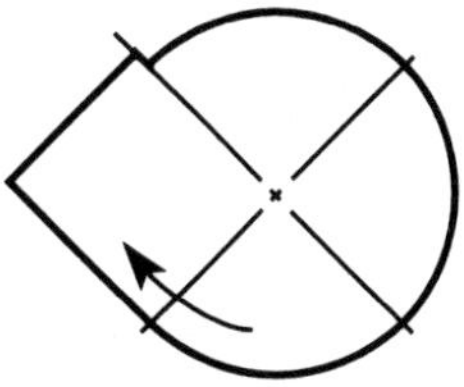

17. ____________________

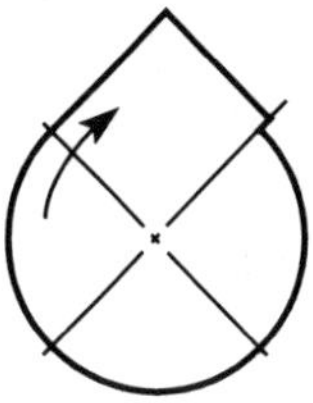

18. ____________________

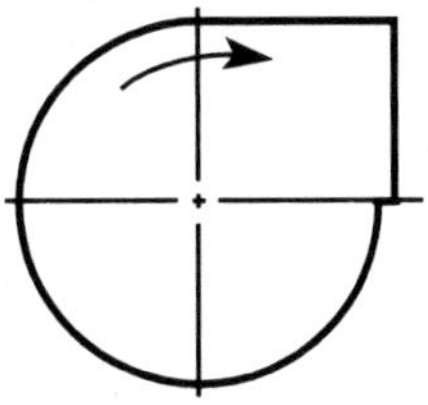

19. ____________________

4 FAN CURVES AND FAN TABLES

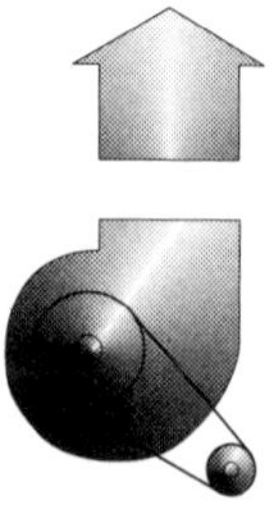

Fan manufacturers publish fan curves and fan tables to assist the HVAC system designer in selecting the proper fan. Although fan curves and tables are basically a design tool, the technician who deals with fans should understand what they indicate and how they are used. The indoor environment technician may find them useful in predicting the results of changes in fan or system operating conditions.

A **fan curve** (Fig. 1) shows the performance characteristics of a fan by means of a graph. A **fan table** shows the same relationships but with specific values in tables. The system designer uses fan curves to see the characteristics of fans but consults fan tables to select a specific model.

See the Glossary in the back of the book to review terms used in this chapter.

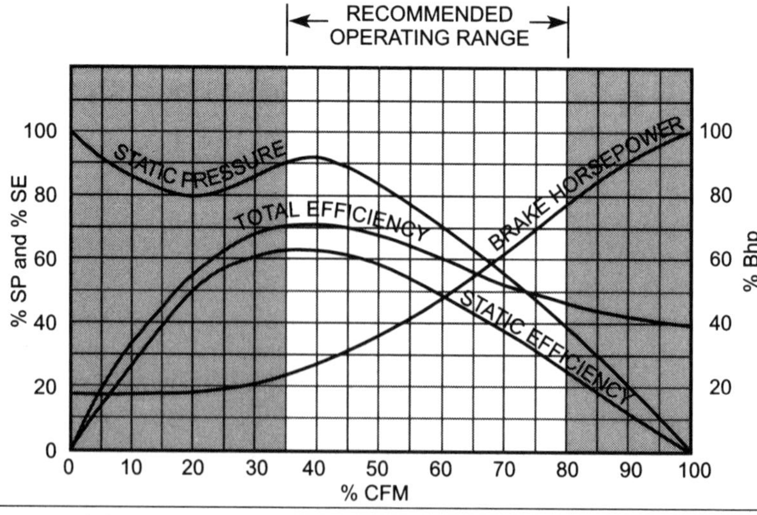

Fig. 1: Typical fan curve for a forward-curved-blade centrifugal fan

FAN CURVES

Many technicians do not understand what a fan curve is. You will have an advantage over most technicians if you understand what it indicates. A **fan curve** (Fig. 1):

- Is a graphic picture of the performance of a fan running at a given speed with various resistances imposed.
- Shows the relationship of SP (static pressure), CFM (cubic feet per minute), Bhp (brake horsepower), and fan SE (static efficiency). It may also show other values such as TE (total efficiency).
- Is developed from the results of a laboratory test under controlled conditions and with a specially designed straight discharge duct. It does **not** indicate what the fan performance will be after a duct system is connected to the fan in the field.

A **general fan curve** (the most common type) shows percentages of values and may apply to several sizes of similar fans. Figures 1, 3, 4, and 5 are general fan curves. A **specific fan curve** shows actual values for a particular model and size at a given RPM. Specific fan curves are not often shown because fan tables are used instead.

Reading a Fan Curve

To understand a fan curve, learn what the scales on the bottom, left, and right of the graph indicate:

- For most fan curves, the **bottom scale** shows CFM. (100% CFM is the wide open volume.)
- The **right hand scale** usually shows Bhp, but this is not always the case.
- For most fan curves, the **left scale** shows SP and usually shows SE (static efficiency). It may show other values, such as TE (total efficiency).

Developing a Fan Curve

The procedure used to develop a fan curve is the same procedure used to develop a fan table. That is why it is useful to understand it.

Figure 2 shows how a fan curve is developed. Under laboratory conditions, a fan with a special straight duct is run at various settings of a discharge damper and readings of CFM, SP, and Bhp are taken:

- Figure 2A: While the fan is run at 100% of capacity (no resistance), the SP, Bhp and CFM are measured. For this particular fan, the CFM is 6000, the Bhp is 3.0, and the SP is 0.0. CFM_1 is marked on the bottom line, and the points for Bhp_1 and SP_1 are plotted on the graph.

  The discharge duct is throttled and the CFM, Bhp, and SP are measured again. CFM_2 is marked on the bottom line and the points for Bhp_2 and SP_2 are plotted.
- Figure 2B: The duct is throttled a number of additional times and these measurements are repeated to locate a series of points.
- Figure 2C: When enough points have been plotted, they are connected and the Bhp and SP curves are drawn.

After these curves are developed, other curves for such values as efficiency are plotted by calculating them with equations. Static efficiency (SE) is often calculated. It is the efficiency of the fan based on the SP that it produces. The system designer considers the SE and SP when choosing a fan for a particular application.

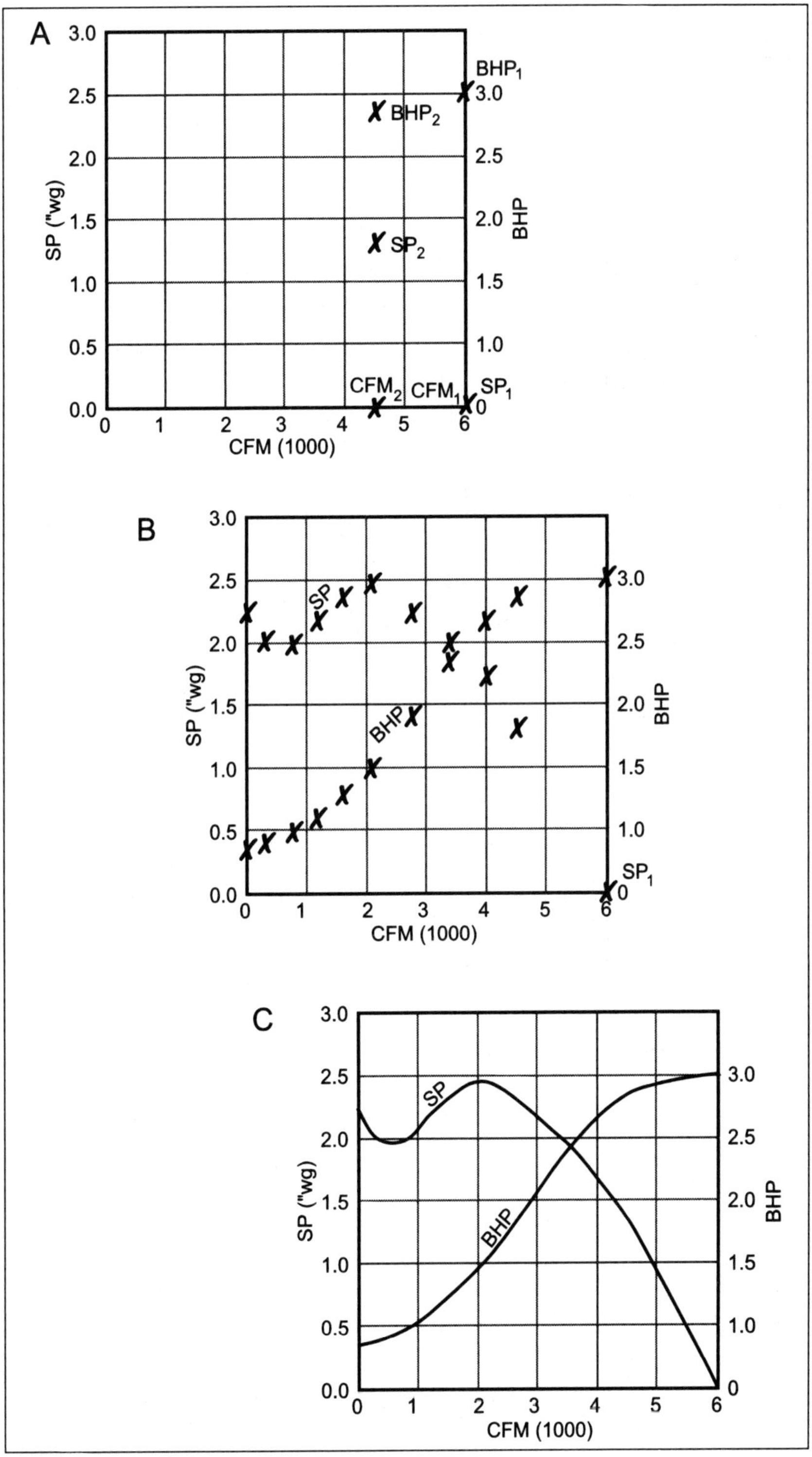

Fig. 2: Developing a fan curve

Figures 1, 3, 4, and 5 show fan curves for fans with different types of fan blades. Each type of fan blade has a different fan curve. However, keep these principles in mind:

- The lower right corner on the bottom scale represents the fan delivering 100% of its capacity. There are no restrictions to airflow. This point is sometimes marked WO (wide open).
- The lower **left** corner represents zero air delivery. The airflow is completely blocked. This point is sometimes marked BT (block tight).
- On all fans, as the CFM decreases from 100%, the SP curve rises, due to added resistance. From zero SP at 100% CFM, the SP will rise to a certain point and then decline. The highest point of SP and the amount of the dip to the left varies with the fan type.
- The SE curve (static efficiency) also rises with the drop in percent of CFM, up to a certain point. Again, the curve varies with the type of fan.
- Most fan curves indicate the **recommended operating range**. In general, the optimum operating range of any fan is at or just to the right of the peak of the SE curve. In general, it is the range that represents the highest SP and SE ranges. If the fan is operating to the left of the recommended operating range, it is oversized, unstable, and less efficient. If the fan is operating to the right of the recommended operating range, it is undersized and less efficient. The least sound power is generated when the fan runs in the recommended range.

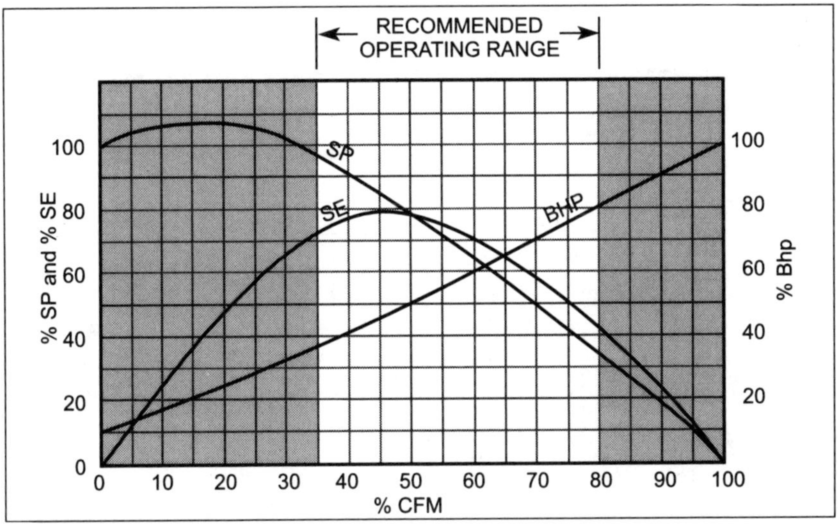

Fig. 3: Radial-blade fan

PRACTICE

To help you see what the fan curve shows, fill out the table below, using the curves in Figs. 1, 3, 4, and 5 as reference. Give the SP, SE, and Bhp readings at 50% CFM. The answers are at the end of this chapter.

	Forward-Curved	Radial-Blade	Backward-Inclined	Vaneaxial
Recommended operating range	1.	6.	11.	16.
SP at 50% CFM	2.	7.	12.	17.
SE at 50% CFM	3.	8.	13.	18.
Bhp at 50% CFM	4.	9.	14.	19.
In the operating range, does Bhp rise or fall as CFM decreases?	5.	10.	15.	20.

RIDING THE CURVE

When VAV (variable air volume) duct systems are discussed, the term "**riding the curve**" is sometimes used. As different VAV boxes open and close, the system SP rises and falls as the system resistance changes. As the SP

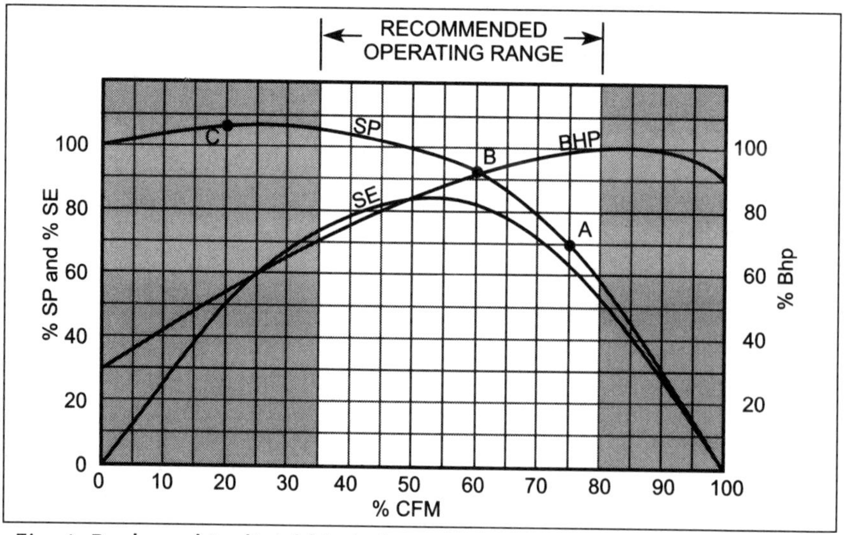

Fig. 4: Backward-inclined-blade fan

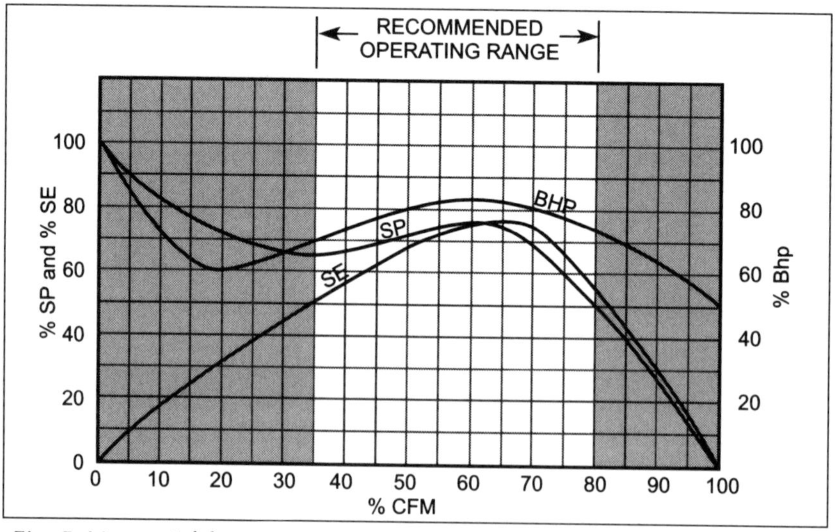

Fig. 5: Vaneaxial fan

imposed on the fan changes, the point of operation moves on the curve, and the fan CFM also changes. Note that when CFM is zero (blocktight) the SP is 100%. Since the fan performance is always on the SP curve, the fan performance will change as the SP "rides the curve."

For example, assume that Fig. 4 is a fan curve for a VAV system fan. The fan is operating at point A, which means that it is 75% of wide open CFM. If the dampers in a number of VAV boxes in the system close at the same time, the system SP increases. This means that the fan operating point rides up the curve and shifts to the left. When it rises to point B, the CFM has dropped to 60%, the SE has increased to 82%, and the Bhp has decreased from 100% to 92%.

If dampers in the VAV boxes continue to close, the fan continues to ride the SP curve. When it reaches point C, the fan is out of the recommended operating range. The result is high air noise, an unstable system, and low fan efficiency.

This illustrates why newer VAV systems use other methods for controlling the system CFM, such as varying the fan RPM.

FAN TABLES

A **fan table** (Fig. 6) contains the same data as a fan curve. The difference is that the table provides specific values for many different RPMs while a curve represents only one RPM. The technician may use a fan table to predict fan performance after fan RPM is changed.

The term **OV** (**outlet velocity**) is used in fan tables (in Fig. 6, at the top of the second column from the left). The OV, stated in FPM (feet per minute) to the nearest 100, is basic information for the fan table. A system designer selects a fan for the proper OV in relation to the fan SP. The proper relationship between OV and SP results in the most efficient and quiet fan.

Fan tables usually indicate the class of fan for a given RPM and OV. (Fan classes are described in Chapter 3.) The table is usually colored or shaded to indicate the various construction classes. In Fig. 6 the shaded areas indicate Class II and Class III fans.

The fan table in Fig. 6 shows that a Class I fan operating at 5.5"wg static pressure must not be speeded up above 2228 RPM. Above that RPM the fan exceeds the manufacturer's limits and is in danger of failing. When an indoor environment technician needs to increase the fan speed in order to obtain more CFM, the fan table will show whether that fan is safe at the necessary speed or if it must be replaced with a fan constructed for a higher fan class. It is always advisable to consult the fan manufacturer's representative before drastically increasing the RPM of a fan.

The indoor environment technician who modifies the system to change the CFM of the fan may find the fan table useful in predicting the conditions after the changes are made. However, the fan table data are obtained under laboratory conditions. They do not reflect the system effect (negative fan performance) that results when the fan is connected to a duct system. In the field, true fan performance (SP, CFM, and Bhp) must be determined by actual measurement.

FAN SURGE

Fan surge means that the CFM and SP are fluctuating in a fan and in the ductwork. The result is noise and vibration in the system, and a pulsing of the air delivered to the spaces.

Fan surge occurs when the CFM drops to a point where the design SP difference between the fan inlet and outlet cannot be maintained. When this happens, the duct static pressure drops and air from the duct surges back into the fan. This surge of air allows the fan to regain its design SP and design CFM. This cycle is constantly repeated, thus producing surges of air.

AIRFOIL BLADE FAN WITH 18" WHEEL

CFM	OV	2¼" SP		2½" SP		3" SP		3½" SP		4" SP		4½" SP		5" SP		5½" SP		6" SP		6½" SP		
		RPM	BHP	RPM	BHP	RPM	BHP	RPM	BHP	RPM	BHP	RPM	BHP	RPM	BHP	RPM	BHP	RPM	BHP	RPM	BHP	
2297	1200	1409	1.16	1475	1.29																	CLASS I
2680	1400	1457	1.34	1515	1.49	1629	1.80	1743	2.11													
3063	1600	1523	1.58	1575	1.73	1678	2.04	1778	2.39	1878	2.74	1978	3.10	2075	3.46							
3446	1800	1597	1.84	1646	2.01	1742	2.36	1834	2.69	1924	3.06	2013	3.45	2102	3.85	2190	4.26	2278	4.66			
3829	2000	1677	2.14	1724	2.32	1814	2.69	1901	3.08	1985	3.45	2067	3.84	2148	4.25	2228	4.69	2308	5.13	2388	5.58	CLASS II
4212	2200	1760	2.47	1805	2.67	1892	3.07	1975	3.48	2054	3.90	2132	4.32	2207	4.73	2282	5.16	2355	5.61	2428	6.09	
4595	2400	1847	2.84	1891	3.06	1974	3.49	2053	3.93	2130	4.37	2203	4.82	2275	5.28	2346	5.73	2415	6.18	2483	6.65	
4978	2600	1937	3.24	1979	3.47	2059	3.95	2135	4.42	2209	4.89	2280	5.37	2349	5.85	2416	6.35	2482	6.84	2547	7.34	
5361	2800	2030	3.68	2069	3.93	2146	4.44	2220	4.95	2291	5.46	2360	5.97	2427	6.48	2492	6.99	2555	7.52	2617	8.06	
5744	3000	2124	4.16	2162	4.43	2236	4.98	2307	5.52	2376	6.07	2443	6.61	2508	7.16	2570	7.70	2632	8.25	2692	8.81	
6127	3200	2221	4.70	2257	4.99	2328	5.56	2397	6.14	2463	6.72	2528	7.31	2591	7.89	2652	8.47	2712	9.05	2770	9.63	
6510	3400	2319	5.30	2354	5.60	2422	6.19	2488	6.81	2553	7.43	2615	8.05	2676	8.67	2736	9.28	2794	9.90	2851	10.50	
6893	3600	2419	5.96	2452	6.27	2518	6.89	2582	7.53	2644	8.18	2705	8.84	2764	9.49	2822	10.10	2878	10.80	2934	11.40	CLASS III
7276	3800	2520	6.69	2552	7.01	2615	7.65	2677	8.31	2737	8.99	2796	9.68	2854	10.30	2910	11.00	2965	11.70	3019	12.40	
7659	4000	2622	7.49	2653	7.81	2714	8.48	2774	9.17	2832	9.87	2889	10.50	2945	11.30	3000	12.00	3053	12.70	3106	13.50	
8042	4200	2726	8.36	2756	8.70	2815	9.38	2872	10.00	2928	10.80	2984	11.50	3038	12.30	3091	13.00	3143	13.80	3194	14.60	

Fig. 6: Typical fan table

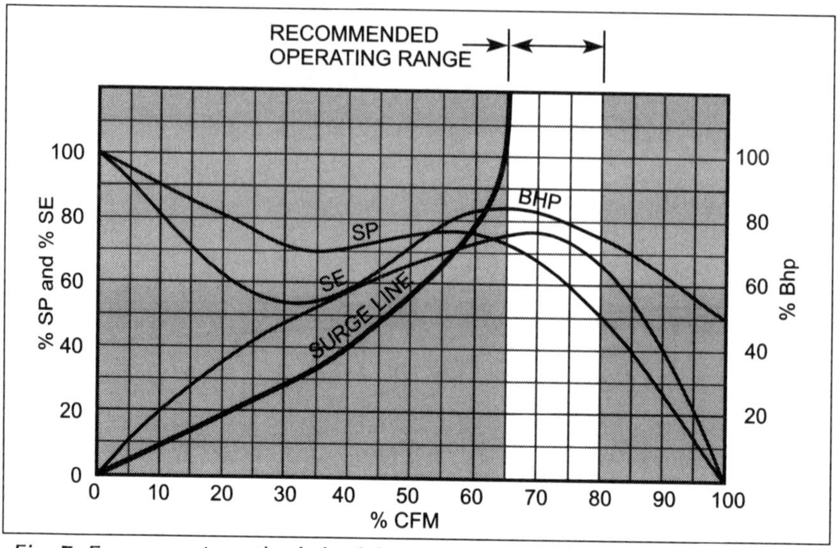

Fig. 7: Fan surge is to the left of the recommended operating range

Fan surge is usually caused by an oversized fan that operates to the left of the static pressure curve. It is also caused by excessive resistance in the system or by poor fan inlet conditions. Some fan curves have a **surge line** (Fig. 7) to indicate conditions that would cause fan surge. In general, the surge line is just to the left of the recommended operating range for the fan.

REVIEW

Use the fan table in Fig. 6 to answer the following questions for standard conditions.

A fan is operating at 4"wg SP and has an OV of 2,800 FPM.

1. What is the Bhp? ____________________

2. What is its RPM? ____________________

3. What is its CFM? ____________________

4. What construction class is this fan? ______________

The fan for a system must operate at an SP of 5"wg and deliver 6,510 CFM:

5. What is its Bhp? ______________

6. What is its OV? ______________

7. What is its RPM? ______________

8. What construction class should this fan be?

____

PRACTICE ANSWERS

	Forward-Curved	Radial-Blade	Backward-Inclined	Vaneaxial
Recommended operating range	1. 35 - 80%	6. 35 - 80%	11. 35 - 80%	16. 35 - 80%
SP at 50% CFM	2. 84%	7. 79%	12. 100%	17. 72%
SE at 50% CFM	3. 58%	8. 78%	13. 84%	18. 68%
Bhp at 50% CFM	4. 37%	9. 50%	14. 84%	19. 80%
In the operating range, does Bhp rise or fall as CFM decreases?	5. Falls	10. Falls	15. Falls	20. Rises & Falls

5 SYSTEM EFFECT

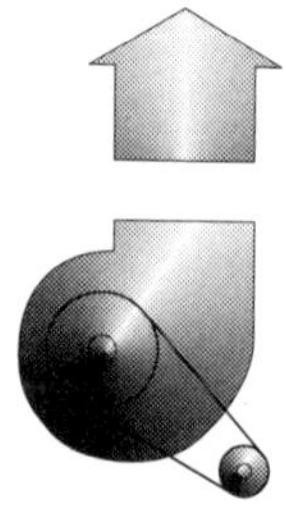

System effect is the negative impact on fan performance that results when the fan is connected to a duct system. Fan curves and fan tables (described in Chapter 4) are developed by laboratory measurements with the fan installed under ideal conditions. However, when the duct system is connected to the fan, the fan operating conditions are changed from those of the laboratory and the fan will have a different performance than as rated.

DESIGNING A SYSTEM

An engineer designs a duct system to take system effect into account and to avoid as much system effect as possible.

The designer first calculates the **system resistance**. This is the amount of resistance to airflow (static pressure) that the system will impose, measured in inches of water gage ("wg). The system resistance depends upon these factors:

- ❏ Friction loss (resistance to airflow caused by duct size, roughness of duct walls, and air velocity)
- ❏ Dynamic loss (resistance to airflow caused by changes in air velocity and direction)
- ❏ Pressure loss (resistance to airflow caused by components such as diffusers, coils, and filters)

After the system resistance is calculated, the designer calculates the **system effect factor**. This is based on the configuration of the fan connection to the system as compared to the ideal situation under which the fan was tested. This system effect factor is based on the use of charts and graphs, such as those in the books *Fans and Systems* published by AMCA (Air Movement and Control Association) and *HVAC Systems Duct Design* published by SMACNA (Sheet Metal and Air Conditioning Contractors' National Association).

Based upon the system effect factor, the **equivalent resistance** from system effect is determined. The **equivalent resistance** is added to **system resistance** to obtain the **total system resistance** in "wg. Then the fan is selected from a fan table, based upon the total system resistance, CFM, outlet velocity, and brake horsepower.

DUCT AT FAN OUTLET

The design of the duct at the fan outlet has a great effect on the system performance. Outlet duct is the concern not only of the system designer but also of the technician who may have to modify the system after it is installed.

For the most efficient fan performance, the duct at the fan outlet should be straight and the same size as the fan outlet. It should be long enough so that the air velocity becomes uniform across the face of the duct. Achieving a uniform air velocity in the duct involves a process commonly called **static regain**.

Static Regain

Static regain is the process of converting velocity pressure (VP) to static pressure (SP). Remember that the total pressure (TP) in a duct system is the sum of velocity pressure plus static pressure (TP = VP + SP). Therefore, for a given TP, if the VP is high (usually because the duct area is small), then the static pressure (SP) must be low.

SP is the pressure that **causes** the air in the duct to flow, and VP is the pressure that **results** from the air movement. This means that it is desirable to have a high value of static pressure (SP) compared to the total pressure (TP) developed by the fan.

Figure 1 shows the air velocity profiles in a duct at various distances from the outlet of a centrifugal fan. The air in the fan is pushed against the outside of the scroll by the movement of the fan wheel. Therefore, at the fan outlet, there is a high velocity at the top of the fan outlet. However, at the bottom of the fan outlet, there is a negative velocity, because the air is swirling back to the fan at the cut-off, attempting to re-enter the fan.

At point A in Fig. 1, the VP is high and the available SP is low. As the air moves down the duct, the velocity of the air becomes more uniform across the duct, and the static pressure increases as the velocity pressure decreases. At point B in Fig. 1 the air velocity is uniform across the duct and low compared to the outlet velocity (point A).

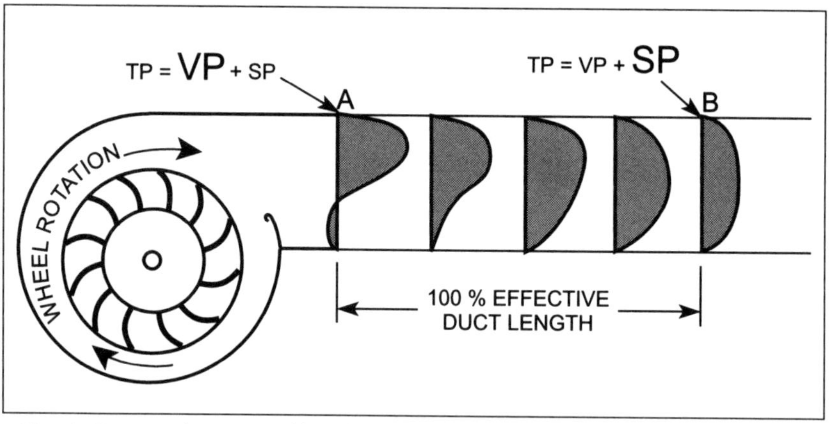

Fig. 1: Duct velocity profiles showing static regain

Remember that TP = VP + SP. Since the total pressure (TP) in the duct at point B is about the same as it was at point A, as the VP has decreased, the SP has increased. In other words, the system has gained static pressure. This is **static regain**. The system now has more potential to overcome the resistance in the system and thus the system can deliver more air.

100%-Effective Duct Length

At point B in Fig. 1, the air velocity is uniform across the duct area and has slowed. This is the point of highest static regain. The distance from A to B is called the **100%-effective duct length**. If possible, the fan outlet should be designed with straight duct for the 100% effective duct length in order to eliminate system effect at the outlet. The technician should try to maintain straight duct at the outlet. If possible, avoid putting a fitting near the fan outlet.

Calculating the 100%-effective duct length depends upon the air velocity at the fan outlet:

- ❑ If the outlet velocity is **less** than 2500 FPM:

  100%-effective duct length = 2.5 x Duct diameter

- ❑ If the outlet velocity is **more** than 2500 FPM:

$$100\%\text{-effective duct length} = \frac{\text{FPM}}{1000} \times \text{Duct diameter}$$

The 100%-effective duct length is always given in **equivalent duct diameters**. However, centrifugal fan outlets are almost always rectangular. To find the equivalent diameter for a rectangular duct of specific dimensions, consult a table (Fig. 2) or a duct calculator (Fig. 3). An **equivalent diameter** is the diameter of round duct that will handle the same amount of air (CFM) at the same friction loss as a given size of rectangular duct. Note that these are NOT **equal areas**—they have **equal resistance** to airflow. (Rectangular and round duct with equal areas have different perimeters and thus offer different resistance to airflow.)

Side Rectangular Duct	6	7	8	9	10	11	12	13	14	15	16	17	18	19	20	22
6	6.6															
7	7.1	7.7														
8	7.6	8.2	8.7													
9	8.0	8.7	9.3	9.8												
10	8.4	9.1	9.8	10.4	10.9											
11	8.8	9.5	10.2	10.9	11.5	12.0										
12	9.1	9.9	10.7	11.3	12.0	12.6	13.1									
13	9.5	10.3	11.1	11.8	12.4	13.1	13.7	14.2								
14	9.8	10.7	11.5	12.2	12.9	13.5	14.2	14.7	15.3							
15	10.1	11.0	11.8	12.6	13.3	14.0	14.6	15.3	15.8	16.4						
16	10.4	11.3	12.2	13.0	13.7	14.4	15.1	15.7	16.4	16.9	17.5					
17	10.7	11.6	12.5	13.4	14.1	14.9	15.6	16.2	16.8	17.4	18.0	18.6				
18	11.0	11.9	12.9	13.7	14.5	15.3	16.0	16.7	17.3	17.9	18.5	19.1	19.7			
19	11.2	12.2	13.2	14.1	14.9	15.7	16.4	17.1	17.8	18.4	19.0	19.6	20.2	20.8		
20	11.5	12.5	13.5	14.4	15.2	16.0	16.8	17.5	18.2	18.9	19.5	20.1	20.7	21.3	21.9	

Fig. 2: Partial equivalent diameter table

Figure 2 shows only a part of a table for equivalent duct diameters. For a full table for duct up to 90 x 88, see the SMACNA publication *HVAC Systems and Duct Design*. To use the table (Fig. 2), locate one of the duct dimensions in the column on the left, and the other duct dimension in the row at the top. The intersection of the vertical and horizontal columns shows the equivalent diameter.

For example, to find the equivalent diameter of a 14" x 12" duct in the table in Fig. 2:

1. Locate 14 in the column on the left.
2. Locate 12 in the row on the top.
3. The intersection of the vertical column and the horizontal row shows 14.2, so the equivalent round duct for a 14" x 12" duct is 14.2" diameter.

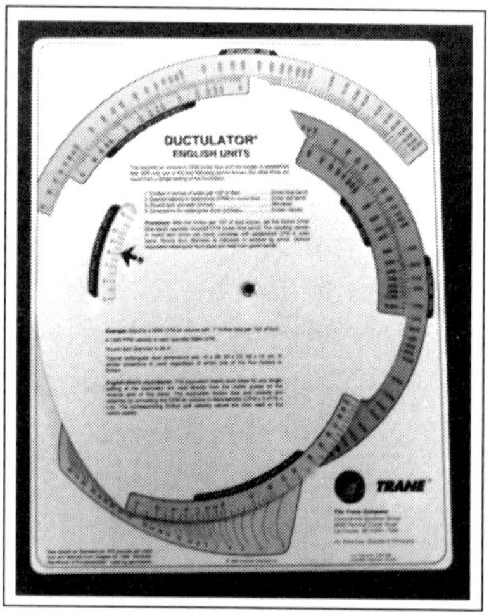

Fig. 3: Duct calculator

A duct calculator (Fig. 3) works in much the same way. Align the marks for two duct dimensions for rectangular duct. Another scale will indicate the equivalent round duct diameter.

Example:

What is the 100%-effective duct length for a fan outlet that measures 20" x 14" if the air velocity is 3500 FPM?

1. The table in Fig. 2 shows that the equivalent duct diameter for a 20" x 14" duct is 18.2".
2. Calculate the 100%-effective duct length:

$$100\%\text{-effective duct length} = \frac{\text{FPM}}{1000} \times \text{Duct diameter}$$

$$100\%\text{-effective duct length} = \frac{3500}{1000} \times 18.2''$$

$$100\%\text{-effective duct length} = 63.7''$$

Straight Duct Length

Even if a 100%-effective duct length is not possible, make the straight duct as long as possible. The closer it is to 100%-effective duct length, the more static regain occurs. The table in Fig. 4 illustrates the percentage of static regain for a centrifugal fan at different percentages of effective duct length. For example, at 50% of the effective duct length, the duct has 80% of the static regain.

	Percent of effective duct length				
	No Duct	12%	50%	90%	100%
Percent of static regain	0%	50%	80%	90%	100%

Fig. 4: Relation of static regain to effective duct length

CAUSES OF SYSTEM EFFECT

Even with the most careful calculations, the installed fan will seldom perform exactly according to the design specifications because of system effect (the negative impact on fan performance caused by the duct system). Many unpredictable conditions occur during the system fabrication and installation.

Generally, system effect is the result of less than ideal conditions at the fan inlet or outlet. The four most common causes of system effect on the fan are:

- ❑ Uneven airflow into the fan inlet
- ❑ Spinning air at the fan inlet
- ❑ Obstructions at the fan inlet or outlet
- ❑ Improper duct at the fan outlet or inlet

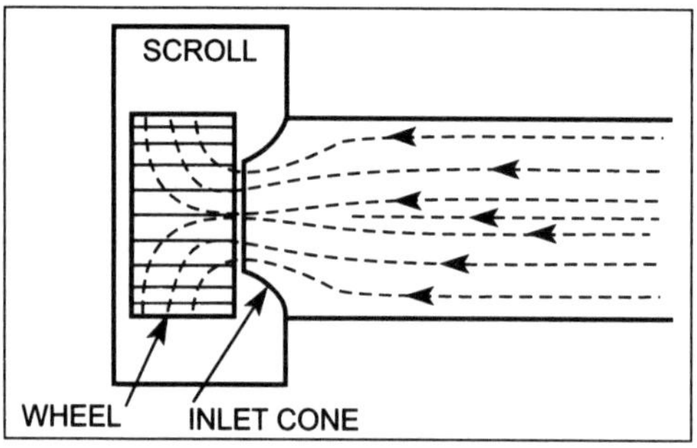

Fig. 5: Air should enter fan inlet evenly

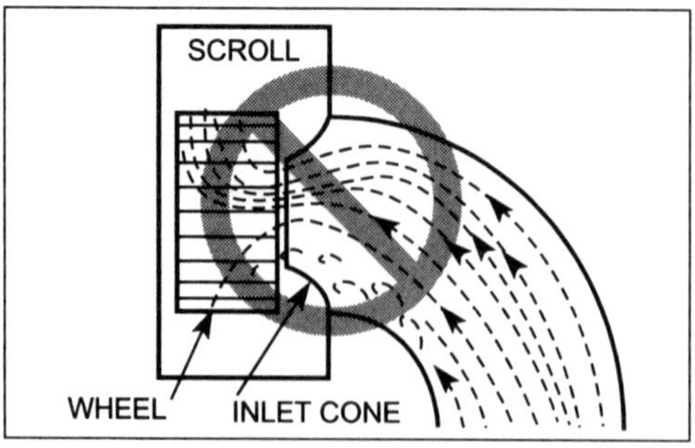

Fig. 6: Uneven air entry into fan

Uneven Flow at Fan Inlet

For best fan performance, air should enter the fan inlet at the **same velocity** across the entire inlet area (Fig. 5). With these conditions, all the fan blades can handle the maximum amount of air.

Uneven distribution of air velocity at the inlet (Fig. 6) causes turbulence in the fan and less air delivery.

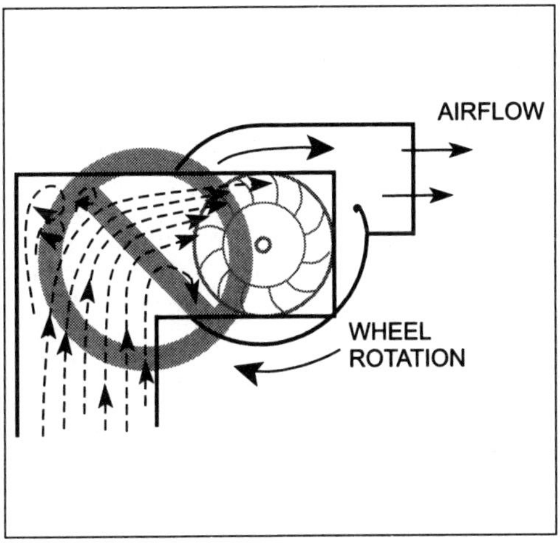

Fig. 7: Air spin with fan wheel rotation

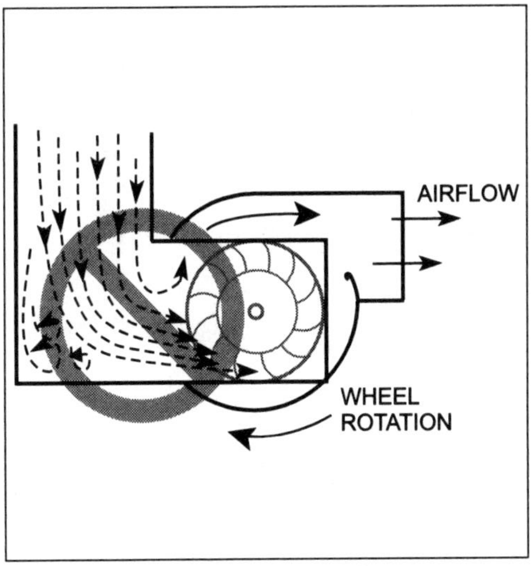

Fig. 8: Air spin against wheel rotation

Spinning Air at Fan Inlet

In addition to the air entering the fan at equal velocity across the inlet area, the air should enter the inlet in a **straight path**. This allows equal distribution of the air to all fan blades. Often improper fan inlet connections cause the air to enter in a spinning motion.

If the entering air spins **in the direction of the fan rotation** (Fig. 7), the pressure and volume produced by the fan is reduced.

If the spin is **opposite the fan rotation** (Fig. 8), the air volume and pressure may be increased. However, there will also be an increase in the required brake horsepower that is out of proportion. The cost of energy because of the increased horsepower makes it economically impractical. In addition there will be undesirable air noise.

Obstructions at Inlet or Outlet

Anything that hinders the airflow at the fan outlet or inlet will have a system effect. Sound attenuation material in the duct that is installed too close to the fan outlet can be blown loose and obstruct the airflow. A fan inlet that is too close to the plenum wall can have restricted airflow. Material in the fan plenum that does not belong there (such as ladders, boxes, or rolls of insulation) can cause turbulence and restricted airflow that will result in system effect.

Proper Outlet Connections

On most jobs there is not enough space at the fan outlet to install a straight duct that is equal to a 100%-effective duct length. However the duct designer and the duct installer must do everything possible to keep the system effect as low as possible. Follow these rules when connecting duct to a fan outlet:

- ❏ Use as long a straight duct as possible for the job conditions.
- ❏ If a duct transition must be used, slant the sides as gradually as possible.

Gradual Transition (Evasé) at Outlet

If the performance of a fan is deficient, it may be possible to install an **evasé** (ev-a-say') designed by the manufacturer for the particular fan. An evasé is a duct transition with gradually slanted sides and bottom that connects to the fan outlet and enlarges the duct to the size of the supply duct. As the duct area of the transition increases, the air velocity (FPM) slows and becomes uniform across the transition area, achieving static regain.

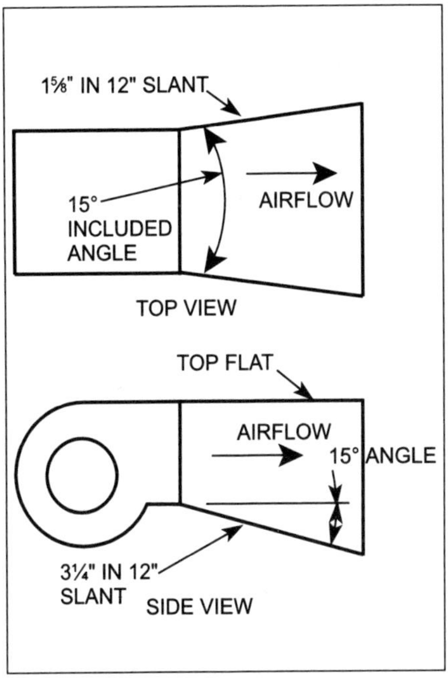

Fig. 9: Usual design for an evasé

With a properly designed and fabricated evasé, the pressure recovery will be the same as with a 100%-effective duct length. The design of the evasé depends upon the fan design and the fan operating conditions. Therefore, for full effect, the design information should be obtained from the fan manufacturer.

The evasé is usually flat on the top and slants very gradually on the sides and bottom (Fig. 9). In general, the included angle for the sides is 15° and the bottom slants down 15° (Fig. 9). If you cannot install a full evasé, do what you can in the space available.

Proper Inlet Connections

For maximum performance, the air should enter straight into the fan inlet, with a uniform velocity across the area of the inlet. The ideal inlet connection is a long, straight duct with a length four times the diameter of the inlet. If an elbow is required, there should be a length of straight duct between the fan inlet and the elbow at least two times the diameter of the fan inlet (Fig. 10).

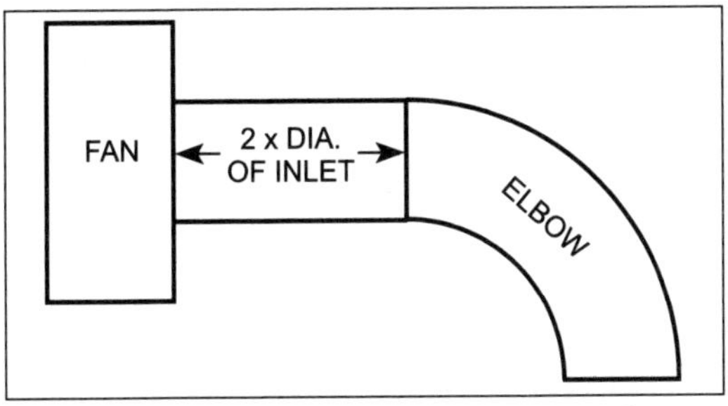

Fig. 10: Proper inlet connection with elbow

Job conditions seldom allow enough space for the ideal inlet condition. Often elbows must be attached directly to the fan inlet with little or no straight duct. Do what you can to minimize the system effect:

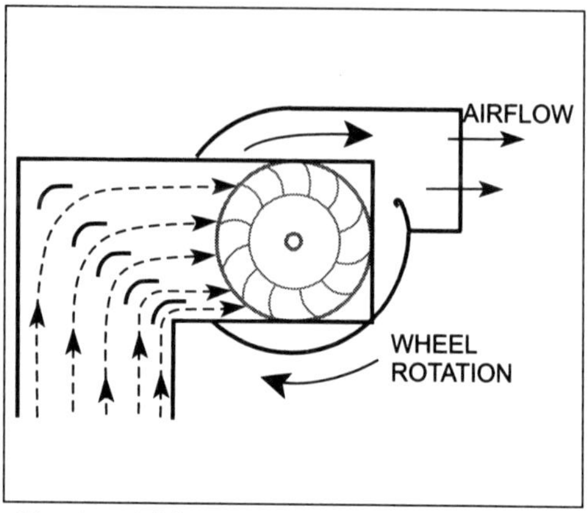

Fig. 11: Add turning vanes to reduce air spin

- ❑ For rectangular duct:
  - ❑ With square throat elbows, use turning vanes (Fig. 11).
  - ❑ With radius throat elbows, make the throat radius as large as the width of the elbow cheek, if possible.
- ❑ For round duct elbows, make the throat radius the same as the duct diameter.

REVIEW

If you can answer the following questions without referring to the text, you have learned the contents of this chapter. Try to answer every question before you check the answers in the back of the book.

1. When a duct system is connected to a fan, it results in reduced fan performance. What is this called?

2. Name three factors that contribute to resistance in a duct system.

____

____

____

3. Resistance in a duct system that results from duct size and air velocity is called ____________________

4. Resistance in a duct system that results from duct fittings is called ________________________

5. Resistance in a duct system that results from system components is called ______________________

6. What device can be used to reduce air spin at the fan inlet?

____

7. What process results from enlarging the duct to slow the air velocity?

____

8. A duct has the equivalent diameter of 12". The air velocity in the duct is 2000 FPM. What is its 100%-effective duct length?

____

9. A duct has the equivalent diameter of 12". The air velocity in the duct is 3500 FPM. What is its 100%-effective duct length?

____

10. What is the equivalent diameter of an 18" x 18" duct?

---

11. What is the name of a special duct fitting at the fan outlet that is designed by the fan manufacturer?

---

6 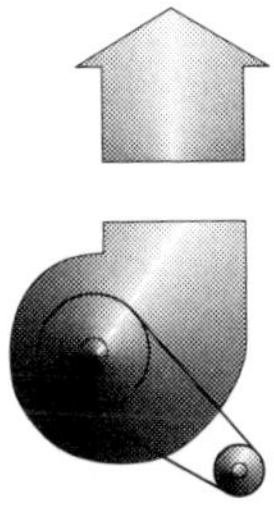

V-BELT DRIVES

V-BELT DRIVE TERMS

Small fans often are **direct drive**. This means that the motor is directly connected to the fan shaft. However almost all residential and commercial centrifugal fans, and some in-line fans, use a **V-belt drive** to transmit power to the fan. The V-belt drive (Fig. 1) consists of:

- ❑ **Drive sheave** (motor sheave) (may be called **drive**)
- ❑ **Driven sheave** (fan sheave)
- ❑ **V-belt**

> In the HVAC industry, *sheave* is pronounced *shiv* (rhymes with *give*).

V-belt drives are required to change the motor RPM to a correct speed for the fan (usually less than the motor speed).

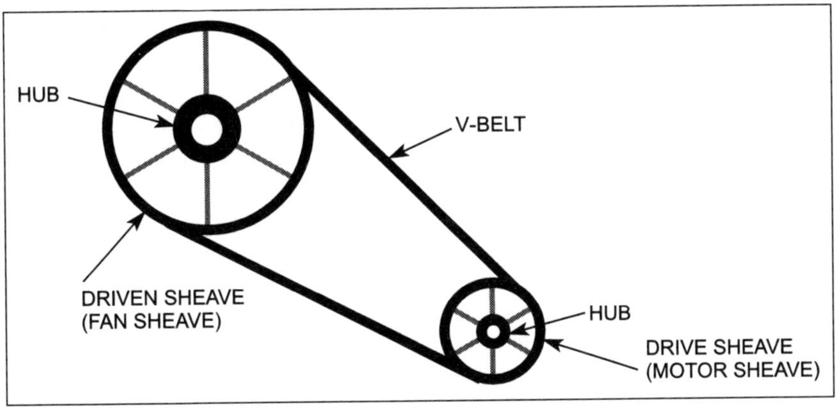

Fig. 1: V-belt drive

Inertia

Inertia is the tendency of an object at rest to remain at rest, or of an object that is moving to continue moving in the same direction. Inertia in a fan at rest means that extra force is required to start a fan turning. Once the fan reaches full speed, because of inertia it requires less force to continue moving.

Torque

Torque is a force transferred by a twisting or rotating motion. There is torque between the motor shaft and the motor sheave, and between the fan shaft and the fan sheave.

Keys and Keyways

Sheaves are commonly secured to the shaft are by means of **keys** and **keyways** (Fig. 2). A **keyway** is a groove in the shaft and in the hub of a sheave or its bushing. The **key** is a square or rectangular piece of hard steel that fits into the keyway. This provides a positive connection to prevent slippage. Sometimes a **set screw** in the hub holds the key in place.

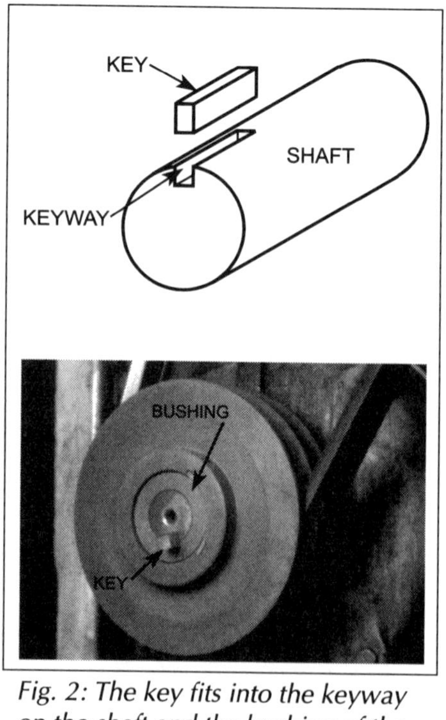

Fig. 2: The key fits into the keyway on the shaft and the bushing of the sheave

Bushings

Bushings (Fig. 2) are used to adapt the hole size of the sheave to the shaft diameter. Sheaves are often manufactured with only one large hole size. For example a sheave may have a bore diameter of $2\frac{7}{8}$" while the fan shaft it is to fit may be $2\frac{1}{16}$" diameter. A bushing is inserted into the hole in the sheave to reduce the bore diameter to the $2\frac{1}{16}$" shaft size. The bushing is usually designed to bolt to the sheave for a positive connection. Because the bushing is tapered, it compresses and makes a tight fit on the fan or motor shaft.

Pitch Diameter, Datum Diameter

Sheave diameters are stated as **pitch diameter** (**PD**) or **datum diameter** (**DD**). The pitch diameter or datum diameter of the sheaves is important because it is the diameter that determines the speed ratio between the motor and the fan.

Pitch diameter used to be the only term used. Now the term **datum diameter** is used, **but only for sheave sizes A, B, C, and D.** (Belt sizes are explained later in this chapter.) The term **pitch diameter** is still commonly used in the industry. However, when you are using catalogs to look up sheave sizes, you have to look for the term datum diameter (**DD**) for A, B, C, and D sheaves.

Pitch diameter (Fig. 3) is different than outside or inside diameter. The pitch diameter or datum diameter is the working (effective) diameter of the sheave. Determine the PD or DD by referring to a catalog. When diameter is stated, it is always assumed to be the pitch diameter or datum diameter unless it is marked OD (outside diameter).

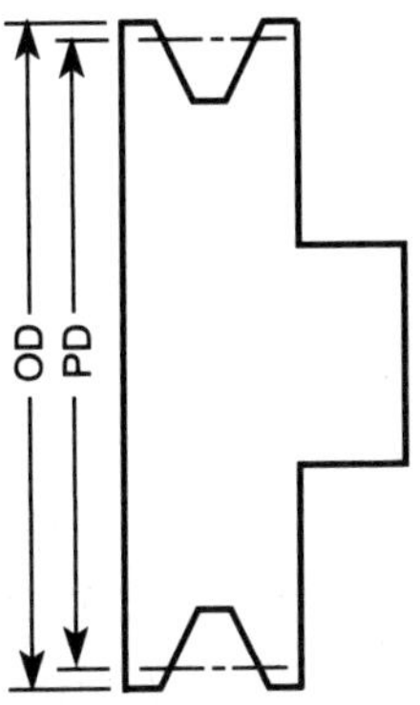

Fig. 3: Pitch diameter and outside diameter

Variable Pitch Sheaves

A **variable pitch sheave** is used for the motor sheave so that one or both of the sheave flanges can be moved to widen or narrow the V-notch. Changing the width of the V-groove makes the V-belt fit higher or lower in the V. This changes the pitch diameter of the sheave. In this way, if the fan speed needs to be changed, the pitch diameter of the sheave can be adjusted instead of replacing it with a different sheave.

There are several different types of variable pitch sheaves. The most common type of variable pitch sheave is shown in Fig. 4. The inside flange is fixed. The outside flange is threaded. The outside flange can be turned on the threaded portion to narrow or widen the V-groove. This moves the V-belt up or down in the V and changes the pitch diameter.

COURTESY EMERSON POWER TRANSMISSION MANUFACTURING

Fig. 4: Variable pitch sheave

To change the pitch diameter of the sheave, remove the V-belt, loosen the set screw on the outside flange, and turn it to adjust the pitch diameter. In general, these sheaves have a range of 1" pitch diameter. In other words, a sheave with a minimum 3.4" PD can be adjusted to a maximum of 4.4" PD.

Variable pitch sheaves are used when the fan speed adjustment will be relatively permanent—the motor sheave PD may never be changed again. If this type of variable pitch sheave is changed to a new PD, both the V-belt alignment and tension must be adjusted when the belt is replaced.

Notched Belts

Notched belts (Fig. 5) have more belt surface contacting the V of the sheave, and they have a special surface for gripping the V. They have a greater horsepower (Hp) rating than

Fig. 5: Notched belt on a sheave

ordinary belts. Compare the two types if each belt is on an 8" PD motor sheave, running at 1750 RPM:

- Standard V-belt—13.61 Hp
- Notched belt—15.89 Hp

Notched belts should never be used with adjustable pitch sheaves. Never replace a notched belt with a standard one.

Fig. 6: Linked belt

Linked Belts

Linked belts (Fig. 6) are made in links so that they can be adjusted to any length. The links can be added or removed with a twist. These belts are available in A, B, and C sizes.

It is good practice for technicians to carry a length of linked belt on the truck for emergencies. Avoid down time for a fan by making a linked belt to length on the site. The proper belt can be installed as soon as you can get one from the shop.

Multiple Belts

For large fans that require more horsepower than one V-belt can transmit, **multiple belts** are used. V-drives with two to four belts are common. However, sheaves are available for up to 10 belts.

Usually the fan manufacturer determines the number and type of belt required. However, the technician may sometimes need to determine the number of belts needed. This equation can be used:

$$\text{No. of belts} = \frac{\text{Motor Hp x Belt service factor}}{\text{Belt Hp rating}}$$

For example, a fan has a 25 Hp motor. The belt chosen has an Hp rating of 14. The belt service factor is 1.5. (**Belt service factor** is the percentage over the rated Hp that a belt can be used at in order to overcome the fan wheel inertia at start-up.) How many V-belts will be needed for the drive system?

$$\text{No. of belts} = \frac{\text{Motor Hp x Belt service factor}}{\text{Belt Hp rating}}$$

$$\text{No. of belts} = \frac{25 \times 1.5}{14}$$

$$\text{No. of belts} = 2.69 \text{ (3 belts needed)}$$

CHANGING SHEAVES AND BELTS

Often the technician has to change the fan RPM in order to change the air volume that the fan will deliver. This means changing either the motor sheave or the fan sheave. If the fan RPM is to be increased, generally the motor sheave is replaced because it is less costly than the larger fan sheave. However, if the fan is to be slowed, consider replacing the fan sheave with a larger diameter so that the belt speed remains the same. The belt Hp would be reduced if a smaller motor sheave were installed.

On the job, you may have to solve problems regarding sheave diameters and fan RPM. For example, you may need to calculate what fan sheave or motor sheave diameter is needed to produce a fan RPM desired. When the fan sheave remains the same:

- If a **smaller** motor sheave diameter is used, the fan speed **decreases**.
- If a **larger** motor sheave diameter is used, the fan speed **increases**.

The ratio of the motor sheave diameter to the fan sheave diameter determines the fan RPM. For example, if the motor sheave diameter is 6.2" and the fan sheave diameter is 12.4", the ratio is 2 to 1 (12.4 ÷ 6.2 = 2). Since the RPM of a fan motor is usually 1750 RPM, with a 2 to 1 ratio, the fan will rotate at 875 RPM (1750 ÷ 2 = 875).

Chapter 8 covers the equations that are used to determine what sheave diameter changes are needed to achieve the desired conditions.

ORDERING V-BELTS

Technicians often have to order V-belts, sometimes to replace worn ones and sometimes to obtain different lengths because the sheave has been changed. The correct belt must be ordered. The wrong type or size of belt can damage equipment, shorten the life of the new belt, and even reduce the power that is being transmitted to the fan.

Generally, if a sheave is changed, new belts must be ordered. However, if only a small change in the size of a sheave is required, it is possible that the same V-belts can be used under two conditions:

- ❑ If the belts show no sign of wear.
- ❑ If the motor can be moved on the slide rails to provide the proper belt tension.

If for any reason the belt cross section and size must be changed, then new sheaves must also be ordered. The V-groove is designed for a specific belt cross section and size and no others will work properly.

If the belts for a multiple belt drive must be replaced, order all new belts that have a marking to ensure that they are matched.

V-belts are classed by their **cross section**, **pitch length**, and **horsepower** (**Hp**). The belt identification (such as A58) is marked on the outside rim of the belt. It indicates the cross section (A) and the nominal pitch length (58). The Hp ratings are given in catalogs for various motor sheave diameters and RPMs.

Cross Sections

The **cross sections** for V-belts that the technician encounters are:

- ❑ **Fractional horsepower** belts are used on smaller fans. They are generally used in residential applications. The belts are light and flexible so that they fit around the smaller sheave diameters.
- ❑ **A**, **B**, **C**, and **D** belts (Fig. 7) are used on medium and large fans in commercial and industrial

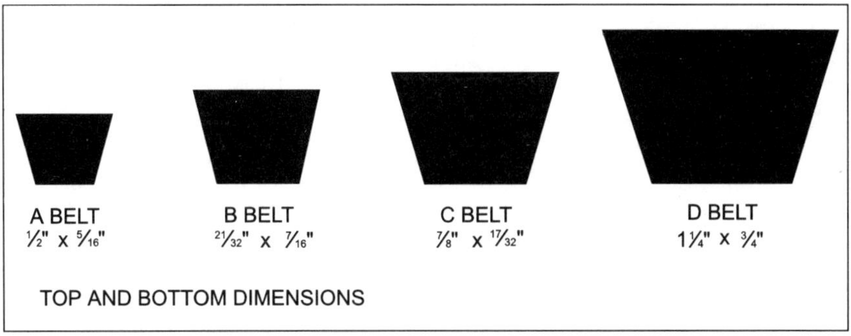

Fig. 7: A, B, C, and D belt sizes

applications. They are designed to transmit the higher horsepowers used in this work. They have an expected life that is generally more than twice that of fractional horsepower belts.

- **3V**, **5V**, and **8V** belts are high capacity belts. These belts must have a higher tension in order to develop their full horsepower. Consequently, they may shorten bearing life.

Each belt section size has a minimum pitch diameter or datum diameter on which it can be used. For example:

Section Size	Minimum DD
A belt	3.0"
B belt	5.4"
C belt	9.0"
D belt	13.0"

Pitch Length

Pitch length is the length of the belt at the point where the belt contacts the pitch diameter of the sheaves.

Belts are usually designated by cross section and **nominal pitch length**. (A nominal measurement is used to identify an item and is an approximate measurement.) For example, a belt listed in a catalog as A50 is an **A** cross section belt with a nominal pitch length of 50". The actual pitch length of an A50 belt is 51.3". Catalogs give the actual pitch length as well as the nominal pitch length.

Horsepower

V-belts are also rated by the **horsepower** (**Hp**) they will transmit. To determine the horsepower of a belt, refer to the manufacturer's catalog.

In general, the Hp rating for a single V-belt should be at least 1.5 times the motor horsepower. For multiple belts, the total of the belt Hp ratings should be 1.5 times the motor horsepower. This extra capacity is to prevent belt slippage when the motor starts and the drive system must overcome the inertia of the fan wheel. If a belt is rated at less than motor Hp, it will slip at starting—and may slip during running. The belt will also have a shorter life and may damage the motor sheave. If you see a motor sheave that is polished in the V, the belt is probably slipping, perhaps because Hp rating of the belt is not adequate.

The horsepower rating of a belt depends upon its cross section, the belt speed, and the PD or DD of the motor sheave. For example, catalog data shows that an A belt with a 4.6" DD motor sheave for 1750 RPM will transmit about 4.9 Hp. The same A belt at the same RPM but with a 6" DD motor sheave will transmit 7.3 Hp. The larger the motor sheave, the faster the belt speed and the greater the Hp capability.

Determining New Belt Length

For convenience, catalogs provide tables so that the technician can choose a belt based on the diameter of the motor sheave and the fan sheave. However, these do not provide all of the possibilities you may encounter, so you should be able to use an equation to calculate the belt length.

There are two equations for calculating the pitch length for a new belt after sheaves have been replaced.

EQUATION 1 for Belt Length

New belt length = 1.5 (New sheave PD – Old sheave PD) + Old belt length

For example, a fan sheave of 12.4" PD has been changed to a fan sheave PD of 15.4". The old belt is size B68. What should the new belt pitch length be and what belt size has this pitch length?

New belt length = 1.5 (New sheave PD – Old sheave PD) + Old belt length

New belt length = 1.5 (15.4" – 12.4") + 68"

New belt length = 72.5" (Use a B71 belt which has a pitch length of 72.8")

From the calculated length you can determine the new belt to be chosen. The catalog indicates the actual pitch length for different nominal pitch lengths.

EQUATION 2 for Belt Length

Equation 2 is not as accurate as equation 1, but it is accurate enough for all practical purposes because belt tension is adjusted by moving the motor.

New belt length = (2 x CD) + 1.5 (D + d)

CD = Center-to-center distance of shafts

D = Pitch diameter of fan sheave

d = Pitch diameter of motor sheave

For example, a motor sheave has been changed to increase the fan speed. The new motor sheave DD is 3.8". The fan sheave DD is 15.4". The CD is 23¼". The old belt is an A section size. What should the new belt pitch length be and what belt size has this pitch length?

New belt length = (2 x CD) + 1.5 (D + d)

New belt length = (2 x 23.25") + 1.5 (15.4" + 3.8")

New belt length = 75.3" (Use an A74 belt which has a pitch length of 75.3")

CHANGING SHEAVES AND BELTS SAFELY

The technician often has to change the fan or motor sheave or change the V-belts. However, you should never make changes to the fan system unless:

- ❑ You know how to calculate the new RPM that results from the change.
- ❑ The resulting power requirements will not overload the motor.
- ❑ You know the RPM limit of the fan. Speeds over the designed limit of the fan can damage the fan wheel.

Whenever you work on a fan drive system, observe these safety rules, which were also covered in Chapter 2:

- ❑ Lock out and tag out—Turn off the main fan disconnect switch, put a padlock on it, and tag it. Even if the switch is already locked out, put your own lock on it.
- ❑ There may also be another switch within reach of the fan. Turn it off and on to be sure the fan motor is de-energized. Sometime a switch is wired wrong

and can still be energized even though the switch is off.

- ❑ If there is no other way to be positive the power is off, test with a voltage meter.
- ❑ Let the fan roll to a stop. NEVER try to stop it with your hands.
- ❑ Block the fan with a piece of wood. Backdrafts can turn the fan wheel.
- ❑ When entering a fan plenum chamber, use a flashlight to check for:
  - ❑ Exposed moving parts
  - ❑ Head hazards (such as low light fixtures)
  - ❑ Location of the light switch
  - ❑ Belt guards in place
- ❑ When entering a fan plenum, block the door open. Under some conditions the air pressure from an operating fan can create enough pressure to make it impossible to open a shut door. If you are working alone, carry a cell phone in order to call for help in an emergency.
- ❑ If you remove a belt guard, replace it as soon as possible. Never leave a fan belt unguarded. It can injure the next person.
- ❑ NEVER put your fingers under a V-belt, even if the fan is off. Any motion of the fan or belts can pull your fingers between the belt and the sheave.

Removing V-Belts

The V-belt must be removed before a sheave can be changed. To do this, loosen the motor mount bolts and move the motor towards the fan sheave to release the tension on the belt. Then roll the belt out of the groove on the fan sheave.

DO NOT PUT YOUR FINGERS AROUND THE BELT. Use the heel of your hand, a piece of wood, or a smooth part of a tool to exert sideways pressure on the belt to roll it out of the groove.

Removing Sheaves

Before removing a sheave, check to be sure that all set screws have been loosened and there are no obstructions on the shaft. If there are burrs, rough spots, or rust on the shaft, sand it lightly and use a fine file to remove any burrs. Then put a light coat of oil on the shaft.

Fig. 8: Extracting bolts on a bushing

Some sheaves have **extracting bolts** (Fig. 8), which are hex head bolts. Tightening these bolts pushes the sheave off the shaft.

If there are no extracting bolts, use a **wheel puller** (Fig. 9) to remove the sheave.

Fig. 9: Wheel puller

On old installations, the hub of the sheave may be frozen on the shaft. If the sheave cannot be moved, apply **penetrating oil** and let it soak long enough (overnight if possible) to free the sheave from the shaft.

As a last resort, hold a block of wood against the **hub only** and tap on the wood lightly. Tapping against the rim of the sheave can break it or knock it out of alignment. Move the wood around the hub so all sides receive the pressure. Do not pound hard because it can damage the fan bearings. NEVER use a steel hammer directly on the sheave to remove or install it. You will damage the shaft or the sheave.

Replacing V-belts

When the new sheave has been placed on the shaft, align it by eye with the old sheave. Be sure keys and bushings are in place and tighten the set screws that secure it to the shaft. These set screws will have to be loosened after the belt is in place in order to do the final belt alignment.

Next, install the belt. The motor should be moved forward toward the fan sheave so that the new belt will have slack. This makes it easier to install the belt. Place the belt into the groove of the fan sheave first and then roll it onto the motor sheave, being careful not to pinch your fingers:

- ❑ **Do not put your fingers on the underside of the belt.** Use the heel of your hand to put pressure against the side of the belt.
- ❑ The belt is always rolled onto the motor sheave.
- ❑ Always roll the belt in the direction of the fan rotation. This places the greatest tension on the leg of the belt that is pulling the fan sheave around. This is especially important on multiple belt drives because it ensures that all the belts have tension on the same leg.

When the belt is in place, move the motor back until the belt has the proper tension. A rule of thumb for belt tension is that a moderate pressure with one finger in the center of the belt on the slack leg should depress the belt about ½".

Do the final alignment of the belts and sheaves. Use a cord or a straight edge (Fig. 10) to check to see if the sheaves are

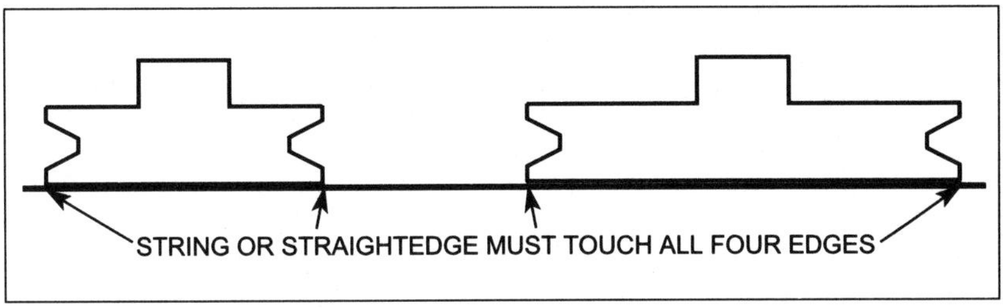

Fig. 10: Align sheave with string or straightedge

properly aligned with each other and are not at an angle. Realign the sheaves if necessary. This is done by moving the motor sheave in and out on the shaft. When the sheaves are properly aligned, a string or straight edge that contacts the outside edges of the fan sheave should also contact the outside edges of the motor sheave (Fig. 10). Tighten the sheave to the shaft. Often the sheave will move slightly when it is tightened to the shaft. Allow for this during alignment.

The motor shaft and fan shaft must be parallel. If they are not, there will be excessive wear on the belts and fan bearings. If the sheaves have been aligned properly, the shafts should be parallel. A good check is to see that the motor is square on the motor mount. If measurement shows that one end of the motor is farther back on the mount than the other end, then the two shafts are not parallel.

It is good practice to re-tension the belt in a day or so. New belts will stretch. During tensioning, the motor can get out of alignment. Check that the motor shaft and fan shaft are still parallel.

DETAILED TECHNICAL INFORMATION

For detailed technical information about belts and sheaves, such as belt lengths, Hp ratings of belts, and datum diameter (pitch diameter) of sheaves, check the manufacturer's catalog or web site.

REVIEW

If you can answer the following questions without referring to the text, you have learned the contents of this chapter. Try to answer every question before you check the answers in the back of the book.

Matching

____ 1. Pitch diameter
____ 2. Datum diameter
____ 3. Inertia
____ 4. Pitch length
____ 5. Torque
____ 6. Key
____ 7. Bushing
____ 8. Fractional horsepower belts
____ 9. A, B, C, D belts
____10. 3V, 5V, 8V belts

A. The working diameter of the sheave
B. A twisting force
C. Reduces hole size in a fan sheave
D. Tendency of a body to remain at rest
E. For medium commercial and industrial applications
F. Used on smaller fans
G. Length at PD
H. High capacity belts
I. Applies only to A, B, C, and D belts
J. Keeps fan sheave from turning on the shaft

11. How can you determine V-belt horsepower?

____

12. What is the advantage of linked belts?

____

13. A fan has a 30 Hp motor. The belt to be used has an Hp rating of 14 and a belt service factor of 1.5. How many belts are needed?

---

7 FAN ACCESSORIES

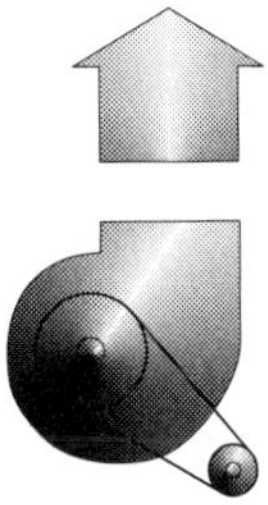

Fan accessories are the equipment associated with a fan to make it a complete installation for its operation. These include such components as:

- Vibration dampening devices
- Belt guards
- Slide rails
- Inlet vanes and outlet dampers
- Variable frequency drives
- Sound attenuation

VIBRATION DAMPENING

No matter how well the fan and motor are installed, there will always be some vibration. Unless this vibration is eliminated at the fan, it will carry through the building structure and through the ductwork and will make the HVAC system unacceptable. There are a number of fan accessories to dampen this vibration. (*Dampen* in this use means to reduce or deaden.)

Vibration Dampeners

Vibration dampeners (also called **isolation blocks**) can be springs or rubber blocks that keep vibration from the fan from carrying to the building.

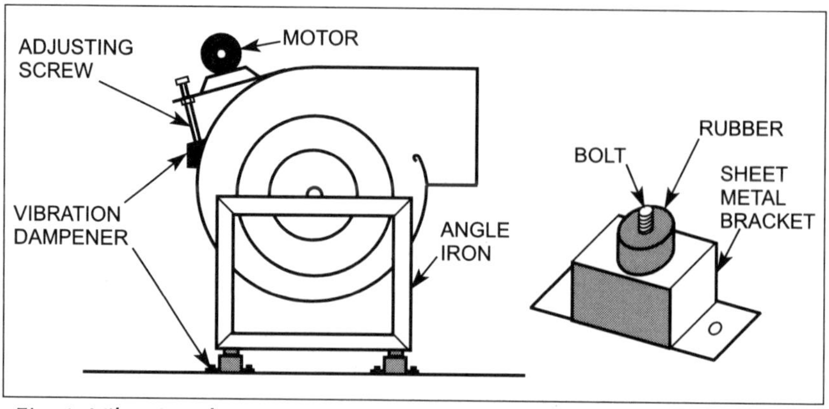

Fig. 1: Vibration dampeners are used on the motor bracket and fan mount

Some vibration dampeners are rubber blocks that have a mounting bolt vulcanized in the rubber. They are used for small fans, such as those used in residential furnaces. The fan motor may be bolted on a bracket on the fan housing with a vibration dampener between the adjusting screw and the fan casing (Fig. 1).

The fan is also mounted on vibration dampeners (Fig. 1). A metal bracket fastens the dampener down (Fig. 1), but this metal does not touch the fan mount, so that no metal from the fan can transfer vibration to the building.

Vibration dampeners for larger fans in non-residential applications are spring mounts under heavy angle iron (Fig. 2). These carry more weight than residential dampeners.

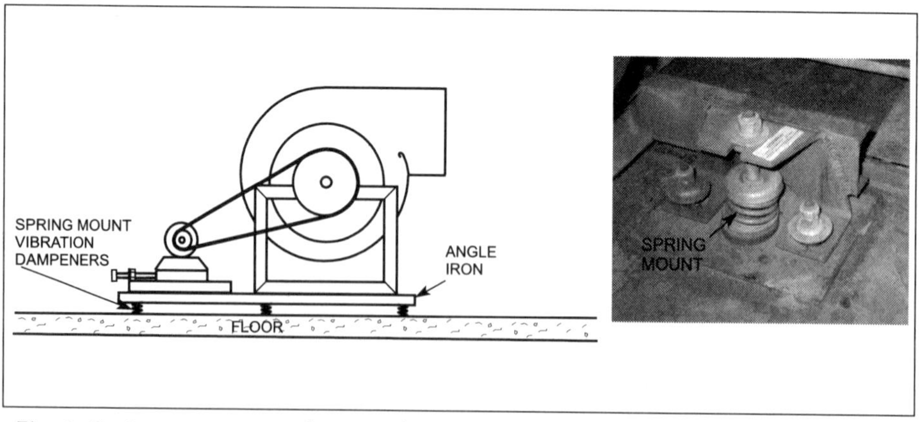

Fig. 2: Spring mounts as vibration dampeners

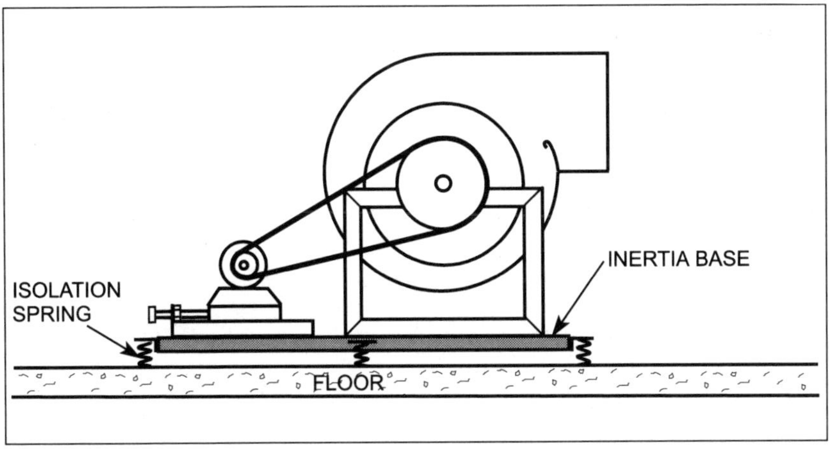

Fig. 3: Inertia base on springs

Inertia Bases

On large installations, the fan and the motor are bolted directly to an **inertia base** (also called an **inertia block**) (Fig. 3). The inertia base is a concrete slab that is attached to the floor by heavy coil springs (Fig. 3). The heavy weight of the inertia base absorbs vibration from the fan and motor, and the isolation springs dampen any remaining vibration to prevent it from carrying to the floor.

The tension on the springs may have to be adjusted, because more compression may be needed at certain points, such as under the motor. If the inertia base is directly on the ground, the springs are not used, since the earth will absorb the vibrations.

In addition to inertia bases, large fans in earthquake zones require **seismic restraints**. These prevent the fan from tipping or moving off its mountings during an earthquake.

Flexible Connectors

If the fan outlet were connected directly to the duct system, the vibration from the fan and motor would be carried by the duct throughout the building. To avoid this, fans are generally connected to the ductwork by a **flexible connector** (Fig. 4). The material of the connector is a tough, fireproof, cloth-like material. This material provides an air-tight connection and none of the fan vibration is carried to the ductwork.

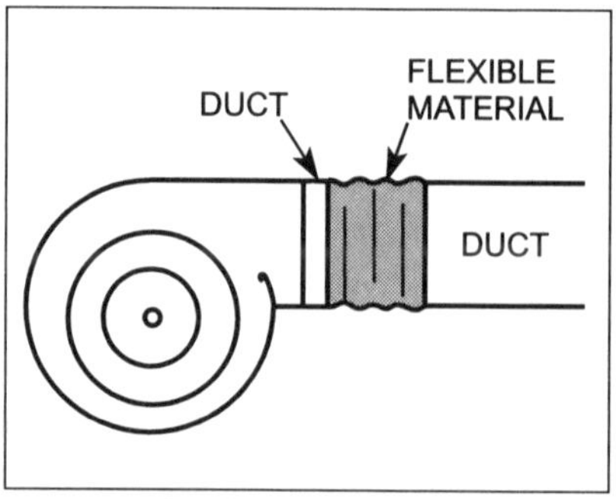

Fig. 4: Flexible connector

Most manufactured flexible connections have a sheet metal strip sealed to both edges of the flexible material. The sheet metal is cut and formed to the size of the fan outlet. Generally, flexible connections are purchased in rolls and cut to length as needed.

Fig. 5: Belt guard

BELT GUARDS

Every fan V-belt drive **must** have a **belt guard** (Fig. 5) to cover all moving parts of the drive. The belt guard blocks some air passage to the fan inlet, and therefore it imposes some system effect on the fan. The front of the guard should be made of screen or mesh to allow as much air passage as possible. Metal brackets hold the guard firmly to the motor and to the fan.

Belt guards are installed for **safety**. Avoid running the fan with the belt guard removed unless absolutely necessary. Even though you are the only person working on the fan, it is very easy to forget about the moving belt and sheaves and get caught by them.

A running V-drive without a guard is also a trap for other workers. They expect the guard to be in place and can be injured very easily. In one case, a worker entering a fan plenum reached inside to steady himself while stepping

Fig. 6: Slide rails and tension adjusting bolts for motor

Fig. 7: Tighten the motor bolts

through the plenum door. The belt guard had been left off and his fingers were smashed between the belt and the sheave.

SLIDE RAILS

Slide rails (Fig. 6) are motor mounts that allow the motor to be slid back and forth to adjust the tension of the belts. The slide rails are bolted to the base and the motor is bolted directly to the slide rail. To move the motor in order to adjust the belt tension, the **motor bolts** are loosened enough to allow the motor to slide. On most mounts, there are **tension adjusting bolts** (Fig. 6) to move the motor forward and back to obtain the proper belt tension, and to align the V-belts. Then the motor bolts are re-tightened (Fig. 7).

If there is only one adjusting bolt, pull the motor back with that bolt. Tighten one bolt that holds the motor to the slide rails. Tighten the adjusting bolt and pivot the motor until the V-belt is aligned. Then tighten all bolts holding the motor.

The slide rails must be bolted to the inertia base so that they are exactly parallel to the fan base. After the belt tension is adjusted, the motor shaft must be at right angles to the slides. This can be checked by measuring that both ends of the motor base are the same distance back from the end of the slide. This ensures that the motor shaft is parallel to the fan shaft.

AIR CONTROLLERS

Air pressures in a duct system can be controlled by changing the fan speed. This can be done by changing the V-belt drives or using a VFD (variable frequency drive). Air pressures can also be varied by modulating an outlet damper or inlet vanes at the fan, but this is the least desirable method of changing the air pressure. It costs more energy than changing the V-belt drive or using a VFD.

Inlet Vanes

Inlet vanes (Fig. 8) can be used to control the volume of air that a fan produces. However, even when full open, inlet vanes impose a negative system effect on the fan performance which must be allowed for by the designer.

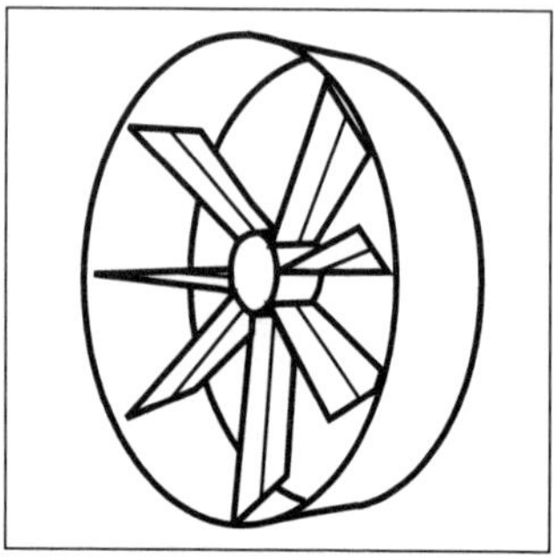

Fig. 8: Inlet vanes

A common application of fan inlet vanes is for variable air volume (VAV) systems. (See the book *Variable Air Volume Systems* in this *Indoor Environment Technician's Library*.) A static pressure (SP) sensor in the supply air duct sends a signal to a controller that opens and closes the vanes as necessary to maintain the proper static pressure in the duct. Inlet vortex vanes are designed to spin the air in the same direction as the fan wheel rotation. On new or retrofit projects, inlet vanes are being replaced by VFDs.

Outlet Dampers

The air volume of a fan can also be controlled by using a damper at the fan outlet. However, even when full open, outlet dampers increase the resistance on the fan. Outlet dampers do not reduce the energy use as much as do inlet vanes, but they are often used because of the low initial cost of installation.

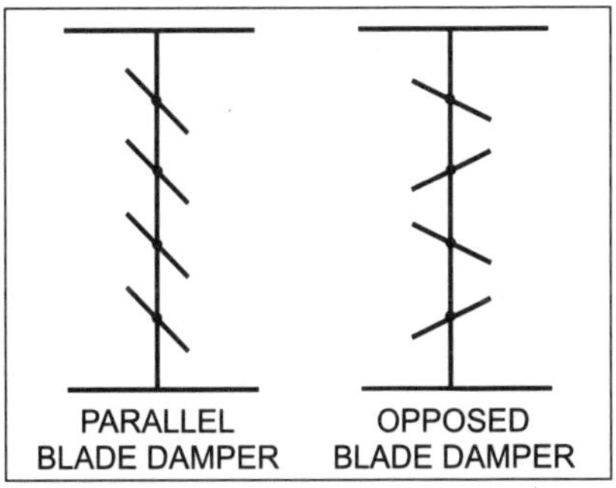

Fig. 9: Dampers may be parallel or opposed blade

There are two types of dampers: **opposed blade** and **parallel blade** action (Fig. 9). A fan outlet damper should be an opposed blade type. This configuration shuts off the air better and provides a more even distribution of air in the duct.

VARIABLE FREQUENCY DRIVES

A **variable frequency drive** (**VFD**) can be used for variable speed control of the fan. A VFD is an electronic device contained in a cabinet (Fig. 10) that controls the motor speed by changing the frequency of the electrical power to the motor. (A VFD is different from the variable pitch sheaves that vary the belt speed in a V-belt drive system.)

Fig. 10: Variable frequency drive (VFD)

Acting on a signal from an outside source, the VFD changes the frequency of the electrical power to the motor. A change in frequency results in a corresponding change in motor RPM and therefore a change in fan speed. In the U.S., the normal frequency of the power supply to a motor is 60 cycles per second (called 60 Hertz or 60Hz). The VFD can vary the frequency to any point between 60 Hz and approximately 12 Hz.

The motor used with a VFD must be designed for that use. If the motor is not designed for a VFD, the fan which cools the motor slows down at lower RPMs. As a result, the motor does not receive enough cooling and will overheat.

One advantage of the VFD is that it can change motor speed automatically, without stopping the motor, in response to changing static pressure in the duct system. This is especially important in variable air volume (VAV) duct systems where the fan speed must be changed frequently to match the changes in the static pressure in the duct.

Another advantage of the VFD is that reducing the motor speed greatly reduces the required energy. This can result in a substantial savings in energy costs. For example, a 25% reduction in fan motor speed could result in as much as a 50% reduction in energy consumption in a VAV system. VFDs automatically improve the power factor, which is beneficial to the owner.

The VFD must be installed and serviced by a specially trained electronic technician only. If it appears that the fan is not operating at the proper speed, the electronic technician should be called in. Although electronic technicians will know how to adjust the VFD, they will probably have no knowledge of duct systems and their requirements. It is your responsibility to see that the VFD is adjusted to meet the needs of the system, and that the fan is not run outside its class.

The VFD is located between the motor disconnect switch and the motor. If the fan motor is to be stopped, turn off the VFD first, before pulling the disconnect switch. If the VFD is running when the disconnect switch is pulled, **the VFD can be damaged**.

Be aware that a VFD can cause **harmonic distortion** in the power supply. This is simply a change in the standard alternating current wave form to the electrical supply to the building. This distortion can affect the operation of other electronic devices in the building, such as computers. If the installation of a VFD affects these other electronic devices, a **harmonic filtering device** can be installed to correct the problem.

Early VFDs tended to overheat at high altitudes. This has been corrected and the VFD is now a highly reliable device.

Depending on the size and model, the VFD can have the following features:

- A soft start, which means that the motor gradually picks up speed until it reaches the desired RPM.
- A digital display that shows the frequency (Hertz), volts, and amps to the motor.
- A manual overide to allow the motor frequency to be temporarily set by hand. This is useful in testing, adjusting, and balancing (TAB) work.
- A motor reverse. This is useful in some cooling tower operations.
- The ability to process either a pneumatic or electronic input signal to the CPU (central processing unit).

SOUND ATTENUATION

Sound attenuation for fans means something designed to lessen the noise made by fans. (*Attenuation* means a lessening of severity.) Not all sound can be eliminated—especially air noises. Therefore many jobs specify that the duct be lined with sound attenuation material. This **duct liner** is a special material fastened to the inside of the duct. It is usually ¾" to 1" thick and is stiff enough to resist tearing from the airflow in the duct. The material is designed to absorb sound.

Often, all straight runs of duct are lined for sound attenuation as well as for thermal insulation. The liner is attached by adhesive and special fasteners.

Duct liner is usually not started at the fan outlet, but is begun after the first duct size increase. At this point, the air velocity has slowed and there is less chance of parts of the liner blowing loose. On a system that has operated for a while, loose material can pile up against turning vanes, reheat coils, and outlets, and seriously affect system performance. Neoprene liners are sometimes used over sound attenuation material to reduce air erosion.

Some installations specify a manufactured **sound attenuator**. This is a device in the duct that deadens noise. These can be very restrictive and can affect the system performance. If fans with manufactured attenuators have poor performance, check the resistance across the inlet and outlet to determine if the attenuator is the cause.

REVIEW

If you can answer the following without referring to the text, you have learned the contents of this chapter. Try to answer every question before you check the answers in the back of the book.

Matching

____ 1. Vibration dampeners
____ 2. Inertia base
____ 3. Flexible connector
____ 4. Belt guard
____ 5. Slide rails
____ 6. VFD
____ 7. Inlet vanes
____ 8. Sound attenuation

A. For adjusting motor position
B. To control volume of inlet air
C. To reduce air noise carried in the duct
D. To vary fan speed electronically
E. To reduce transfer of vibration from a small fan to the building
F. To prevent injury from the V-belt drive
G. To reduce transfer of vibration from a large fan to the building
H. To reduce sound transfer through the metal of the duct

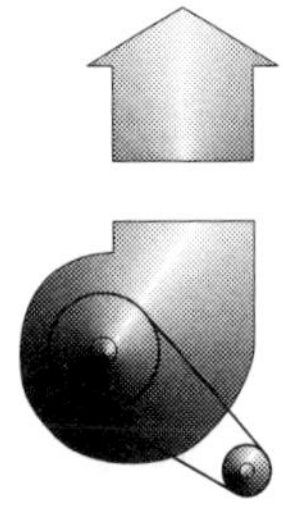

8 FAN LAWS FOR CFM, RPM, AND SHEAVE DIAMETERS

FAN LAWS

Fan laws are equations that relate the performance variables (CFM, RPM, SP, Bhp) of a fan in a fixed system. They compare fan performance before and after a change is made in the operating conditions of a fan. A technician can use fan law equations to determine how to alter the fan operating conditions to obtain the results needed. For example, if the existing RPM and CFM of a fan are known, a fan law can be used to determine the RPM needed in order to have the fan produce a new, required CFM.

Fan Law 1 is used to predict the effect that changing the speed (RPM) has on flow rate (CFM), static pressure (SP), and brake horsepower (Bhp). There are three parts to Fan Law 1:

- **Fan Law 1a** shows the change in CFM with a change in RPM.
- **Fan Law 1b** shows the change in SP with a change in RPM.
- **Fan Law 1c** shows the change in Bhp with a change in RPM.

This chapter explains Fan Law 1a, which deals with CFM and RPM. Chapter 9 explains Fan Law 1b and 1c related to SP and Bhp.

When studying this chapter, you should know the following abbreviations:

- ❑ CFM—Cubic feet per minute
- ❑ CFM_1—Measured CFM; the existing condition
- ❑ CFM_2—New CFM after a particular change is made
- ❑ RPM—Revolutions per minute
- ❑ RPM_1—Measured RPM; the existing condition
- ❑ RPM_2—New RPM after the fan or motor sheave is changed
- ❑ Motor PD—Motor sheave pitch diameter
- ❑ Fan PD—Fan sheave pitch diameter

See Chapter 6 for an explanation of **pitch diameter** (**PD**) and **datum diameter**.

Reminders

1. Increasing the RPM of a fan may exceed its safe mechanical operating conditions. Do not change the RPM of a fan unless you understand its limitations. Consult the manufacturer's fan table or other information.
2. The values determined by fan laws are true only if all conditions in the HVAC air system remain the same. For example, the SP of the fan is calculated for a certain RPM. If some dampers are then closed, the resistance of the system changes and moves the fan performance to a different point on the fan curve. This changes the relationship between fan RPM and SP.
3. This chapter and the following one require the use of ratios, proportions, and equations. If you do not understand these operations, review the book, *Math for the Technician* in this *Indoor Environment Technician Library*, especially the following chapters:

    Chapter 4, Using Equations
    Chapter 5, Rearranging Equations
    Chapter 6, Ratios and Proportions.

    You cannot be a competent HVAC technician if you cannot use and rearrange equations.

FAN LAW 1a

Fan Law 1a is an equation which states that changes in CFM are directly related to changes in the RPM:

$$\frac{CFM_2}{CFM_1} = \frac{RPM_2}{RPM_1}$$

Remember that the subscript 1 indicates the existing CFM or RPM. The subscript 2 indicates the new CFM or RPM after a change is made.

Fan Law 1a means that the CFM and the RPM will change in the same proportion. If the CFM is changed 10%, the RPM will also change 10%.

Like most equations, this equation can be rearranged in different forms. These still state the same relations:

- If you want to determine what the new CFM will be after the RPM is changed, put CFM_2 to the left of the equal sign and the other items to the right:

$$CFM_2 = CFM_1 \times \frac{RPM_2}{RPM_1}$$

- If you want to determine what the new RPM should be to achieve a desired CFM, put RPM_2 to the left of the equal sign and the other items to the right:

$$RPM_2 = RPM_1 \times \frac{CFM_2}{CFM_1}$$

The two rearranged equations again point out the importance of knowing the required math. If you understand how to rearrange equations, you need to remember only the original equation—you will know how to rearrange it to find the value you need.

Example: Find the new CFM

A fan has been running at 450 RPM, supplying 5000 CFM. If the fan speed is changed to 505 RPM, what will the new CFM be?

Known values:

$CFM_1 = 5000$
$RPM_1 = 450$
$RPM_2 = 505$

To find the new CFM, put the term CFM_2 to the left of the equal sign and solve the equation:

$$\frac{CFM_2}{CFM_1} = \frac{RPM_2}{RPM_1}$$

$$CFM_2 = CFM_1 \times \frac{RPM_2}{RPM_1}$$

$$CFM_2 = 5000 \times \frac{505}{450}$$

$$CFM_2 = 5611 \text{ (Round to 5610)}$$

Remember that CFM is rounded off to the nearest 5 in the HVAC industry.

Example: Find the new RPM needed

The measured speed of a fan is 500 RPM, and it has an airflow rate of 1200 CFM. The CFM should be increased to 1400. What should the new fan speed be in order to deliver 1400 CFM?

Known values:

$$RPM_1 = 500$$
$$CFM_1 = 1200$$
$$CFM_2 = 1400$$

To find the new RPM needed, put the term RPM_2 to the left of the equal sign and solve the equation:

$$\frac{CFM_2}{CFM_1} = \frac{RPM_2}{RPM_1}$$

$$RPM_2 = RPM_1 \times \frac{CFM_2}{CFM_1}$$

$$RPM_2 = 500 \times \frac{1400}{1200}$$

$$RPM_2 = 583.33 \text{ (Round to 583)}$$

CHANGING FAN AND MOTOR SHEAVES

Fan Law 1a for Motor Sheave

To achieve a new fan RPM, changes have to be made in the V-belt drive. If a new fan RPM is needed, usually the motor sheave is changed, because it is smaller and cheaper than the fan sheave.

Changes in the motor sheave pitch diameter (Motor PD) are in proportion to changes to the CFM, and changes in the CFM are in proportion to changes in RPM (Fan Law 1a):

- The Motor PD ratio equals the CFM ratio.
- The CFM ratio equals the RPM ratio.

- Therefore the Motor PD ratio equals the RPM ratio. This is a variation of Fan Law 1a:

$$\frac{\text{Motor PD}_2}{\text{Motor PD}_1} = \frac{\text{RPM}_2}{\text{RPM}_1}$$

Use this equation to find the new motor sheave pitch diameter needed. Put the term Motor PD_2 to the left of the equal sign and solve the equation:

$$\text{Motor PD}_2 = \text{Motor PD}_1 \times \frac{\text{RPM}_2}{\text{RPM}_1}$$

Example: Find the motor sheave diameter needed

A fan operating at 1600 RPM has to be speeded up to 1850 RPM. The pitch diameter of the existing motor sheave is 4.2". What should the pitch diameter of the new motor sheave be?

$$\frac{\text{Motor PD}_2}{\text{Motor PD}_1} = \frac{\text{RPM}_2}{\text{RPM}_1}$$

$$\text{Motor PD}_2 = \text{Motor PD}_1 \times \frac{\text{RPM}_2}{\text{RPM}_1}$$

$$\text{Motor PD}_2 = 4.2'' \times \frac{1850}{1600}$$

$$\text{Motor PD}_2 = 4.856''$$

From a manufacturer's catalog, select a motor sheave nearest the pitch diameter of 4.856".

Fan Law 1a for Fan Sheave

In some cases, the fan sheave is changed to achieve a new fan RPM, especially when the RPM is to be reduced. The fan sheave diameter is equal to the **inverse ratio** of the RPM—that is, RPM_2 is the **bottom term** of the ratio and RPM_1 is the **top term** (just the reverse of the RPM term in Fan Law 1a). The equation for the fan sheave is this:

$$\frac{\text{Fan PD}_2}{\text{Fan PD}_1} = \frac{RPM_1}{RPM_2}$$

Note that Fan PD_2 is the **top** term, but RPM_2 is the **bottom** term. Rearrange the equation to solve for the new fan sheave diameter.

How to change sheaves safely is covered in Chapter 6, V-Belt Drives.

Example: Find the fan sheave diameter needed

A fan is operating at 1000 RPM. The RPM must be reduced to 800 RPM. The fan sheave pitch diameter is 20". What must the new fan sheave pitch diameter be?

$$\frac{\text{Fan PD}_2}{\text{Fan PD}_1} = \frac{RPM_1}{RPM_2}$$

$$\text{Fan PD}_2 = \text{Fan PD}_1 \times \frac{RPM_1}{RPM_2}$$

$$\text{Fan PD}_2 = 20'' \times \frac{1000}{800}$$

$$\text{Fan PD}_2 = 25''$$

From a manufacturer's catalog, select a fan sheave nearest the pitch diameter of 25".

REVIEW

These problems give you a chance to apply the knowledge in this chapter.

1. A fan has a speed of 600 RPM and an output of 8800 CFM. The airflow must be increased to 9200 CFM. The motor sheave PD is 5.6".
   A. What will the new speed of this fan have to be in order to produce 9200 CFM (to nearest whole number)?

   B. What will the new calculated motor sheave pitch diameter be (to one decimal place)?

   ---

2. A fan with a speed of 700 RPM produces an airflow of 12,400 CFM. The airflow needs to be increased to 13,000 CFM. The motor sheave PD is 4.6".
   A. What will the new fan speed need to be (to the nearest whole number)?

   B. What will the new calculated motor sheave PD be (to one decimal place)?

   ---

3. The airflow of a fan is 9600 CFM at a speed of 550 RPM. If the speed is increased to 620 RPM, what will the new airflow be (both calculated number and rounded off term)?

   ---

4. If a fan running at 485 RPM produces 8200 CFM, what will the airflow rate be when it runs at 420 RPM (both calculated number and rounded off term)?

   ---

5. For problem 4, the pitch diameter of the motor sheave is 6.4" and the fan sheave is 15.4". The fan sheave will be changed to achieve the new RPM. What will the new calculated fan sheave pitch diameter be?

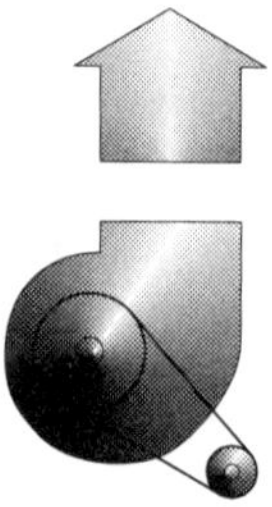

9 FAN LAWS FOR SP AND BHP

Chapter 8 covered Fan Law 1a, which deals with RPM and CFM. This chapter covers Fan Law 1b, which deals with the relationship between fan SP and RPM, and Fan Law 1c, which deals with the relationship between Bhp and fan RPM.

The equations in this chapter involve the math operations of squares, square roots, cubes, and cube roots. **You will need a calculator that has these functions.** If you do not understand squares, square roots, cubes, and cube roots, review Chapter 3 in the book *Math for the Technician* in this *Indoor Environment Technician's Library.*

If you know how to rearrange equations, you do not have to memorize the variations of the basic equations. The process of rearranging and working with equations is explained in Chapters 4, 5, and 6 of *Math for the Technician.*

The reminder in the previous chapter about the danger of increasing the fan RPM beyond its efficient operating range or beyond its mechanical capabilities also applies to the material in this chapter.

FAN LAW 1b

Fan Law 1b shows how the static pressure produced by the fan is related to the RPM of the motor sheave or fan sheave. Fan Law 1b states that the ratio of the static pressures (SP) is in proportion to the **square** of the RPM ratio:

$$\frac{SP_2}{SP_1} = \left(\frac{RPM_2}{RPM_1}\right)^2$$

The equation can be rearranged, based on the **square root** of the SP ratio:

$$\frac{RPM_2}{RPM_1} = \sqrt{\frac{SP_2}{SP_1}}$$

This version of the equation can be changed to determine the new RPM needed if the static pressure must be changed to a new level:

$$RPM_2 = RPM_1 \times \sqrt{\frac{SP_2}{SP_1}}$$

The original equation can also be rearranged to determine what the new static pressure would be if the RPM is changed:

$$SP_2 = SP_1 \times \left(\frac{RPM_2}{RPM_1}\right)^2$$

Example 1: Find the new RPM

The fan SP is measured at 1.8"wg and the RPM is 1050. The fan must operate at 1.6" SP. What will the RPM have to be for the fan to operate at an SP of 1.6"wg?

$$\frac{RPM_2}{RPM_1} = \sqrt{\frac{SP_2}{SP_1}}$$

$$RPM_2 = RPM_1 \times \sqrt{\frac{SP_2}{SP_1}}$$

$$RPM_2 = 1050 \times \sqrt{\frac{1.6}{1.8}}$$

$$RPM_2 = 989.95 \text{ (Round to 990)}$$

The answer is rounded to 990 because RPM is always stated to the nearest whole number.

Example 2: Find the new static pressure

A fan is operating at 1000 RPM. The fan discharge static pressure is 1.5"wg. If the RPM is changed to 1100, what will the new SP be?

$$SP_2 = SP_1 \times \left(\frac{RPM_2}{RPM_1}\right)^2$$

$$SP_2 = 1.5 \times \left(\frac{1100}{1000}\right)^2$$

$$SP_2 = 1.815 \text{ (Round to 1.82"wg)}$$

Determining Airflow Based on SP

By combining Fan Law 1a and Fan Law 1b, you can create an equation to determine the airflow if the static pressure changes. Fan Law 1a states that the ratio of the CFMs equals the ratio of the RPMs:

$$\frac{CFM_2}{CFM_1} = \frac{RPM_2}{RPM_1}$$

Fan Law 1b states that the ratio of the RPM equals the square root of the ratio of the SP:

$$\frac{RPM_2}{RPM_1} = \sqrt{\frac{SP_2}{SP_1}}$$

Since the CFM ratio and the SP ratio both equal the RPM ratio, then they must also equal each other. Therefore:

$$\frac{CFM_2}{CFM_1} = \sqrt{\frac{SP_2}{SP_1}}$$

This equation can be rearranged to solve for the new CFM if the static pressure is changed:

$$CFM_2 = CFM_1 \times \sqrt{\frac{SP_2}{SP_1}}$$

Example 3: Find the new CFM

In a duct system, the airflow in a branch duct was measured to be 2500 CFM and the SP was 0.50"wg. A damper in another branch was restricted, thus increasing the SP in the first branch from 0.50 to 0.65"wg. What is the new airflow in that branch?

$$CFM_2 = CFM_1 \times \sqrt{\frac{SP_2}{SP_1}}$$

$$CFM_2 = 2500 \times \sqrt{\frac{0.65}{0.50}}$$

$$CFM_2 = 2850.4 \text{ (Round to 2850)}$$

FAN LAW 1c

Fan Law 1c shows how the brake horsepower of the fan is related to the RPM of the motor sheave or fan sheave. Fan Law 1c states that the ratio of the old and new brake horsepower (Bhp) is in proportion to the **cube** of the RPM ratio:

$$\frac{Bhp_2}{Bhp_1} = \left(\frac{RPM_2}{RPM_1}\right)^3$$

The equation can be rearranged based on the **cube root** of the Bhp ratio:

$$\frac{RPM_2}{RPM_1} = \sqrt[3]{\frac{Bhp_2}{Bhp_1}}$$

This version of the equation can be changed to determine the new brake horsepower or the new RPM:

$$Bhp_2 = Bhp_1 \times \left(\frac{RPM_2}{RPM_1}\right)^3$$

$$RPM_2 = RPM_1 \times \sqrt[3]{\frac{Bhp_2}{Bhp_1}}$$

Example 4: Find the new Bhp

A fan is operating at 1000 RPM and the Bhp is 20.0. If the RPM is changed to 1100 RPM, what will the new Bhp be?

$$\frac{Bhp_2}{Bhp_1} = \left(\frac{RPM_2}{RPM_1}\right)^3$$

$$Bhp_2 = Bhp_1 \times \left(\frac{RPM_2}{RPM_1}\right)^3$$

$$Bhp_2 = 20 \times \left(\frac{1100}{1000}\right)^3$$

$$Bhp_2 = 26.6$$

COMPARING FAN LAWS 1a, 1b, 1c

Fan Laws 1a, 1b, and 1c all show how the RPM of the sheaves is related to the other variables—airflow (CFM), static pressure (SP) at the fan discharge, and brake horsepower (Bhp) of the fan motor.

These fan laws show that a 10% increase in fan RPM results in:

- ❑ A 10% increase in CFM (Fan Law 1a)
- ❑ A 21% increase in SP (Fan Law 1b)
- ❑ A 33% increase in Bhp (Fan Law 1c)

For instance, Example 2 (page 92) shows that a 10% increase in RPM (from 1000 RPM to 1100 RPM) results in a 21% increase in SP. (An SP of 1.82 is 21% more than an SP of 1.5.)

Example 4 (page 95) shows that a 10% increase in RPM results in a 33% increase in Bhp. (A Bhp of 26.6 is 33% more than a Bhp of 20.)

REVIEW

NOTE: These problems give you a chance to apply the knowledge in this chapter and the previous one.

1. A fan is operating at:
   - ❑ 510 motor sheave RPM
   - ❑ 0.75"wg SP at fan discharge
   - ❑ 11,475 CFM
   - ❑ 2.40 Bhp for fan motor

   The RPM is changed to 630 RPM.

   A. What is the new CFM?

   ____

   B. What is the new SP?

   ____

   C. What is the new Bhp?

   ____

2. A fan is operating at 555 RPM and an SP of 1.25"wg. If the fan RPM is changed and the new SP is measured at 1.0"wg, what is the new RPM?

---

3. A fan is operating at 630 RPM and producing 14,535 CFM. The Bhp is measured at 4.55.

   A. What can the RPM be changed to for the fan to operate at 5 Bhp?

   B. At the new RPM, what will the new CFM be (to the nearest 5)?

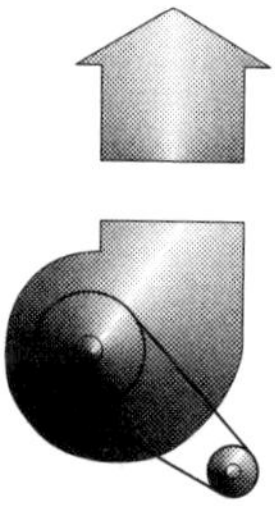

10 FAN PROBLEMS

As a technician, you will be expected to solve fan problems. Most of these problems are relatively simple and can be solved by a common-sense, logical approach.

The basic approach is to identify the source of a problem by a process of elimination. Once the source of the problem is identified, the solution can be determined. For example, if the problem is excessive noise, the first step is to determine if the source is from the fan or from the system. Then use a process of elimination to determine the solution.

GENERAL CAUTIONS

1. To solve fan problems, you will be working on the fan. Remember the basic safety rule: LOCK OUT/TAG OUT. Open the main disconnect switch, put your padlock on it, and tag it with your name and the date. If you are working on electrical circuits, never assume that you know the voltage of the circuit—TEST IT. Even though the fan is locked out and tagged, block the fan wheel with a piece of wood to prevent accidental rotation when you are working on it.

2. Don't go into a fan plenum without ear protection if the fan is noisy.

3. Do not start immediately to take involved instrument readings. Common sense and an industrial stethoscope

can identify many problems. If you do not have a stethoscope, place a screwdriver on the mechanism with your ear at the handle.

4. When working on mechanical systems, remember the "Law of Sixes"— "**Any problem can have six different causes**." (This is something like Murphy's Law that says, "If anything can go wrong—it will.") The point is, if you are troubleshooting and find a problem source, don't assume that you have solved the problem. Check it out. If the problem is still there, it does not necessarily mean that you were wrong. It could mean that there are additional causes of the problem. For example one fan noise problem could be caused by all of the following:

    - ❑ Dirty fan wheel
    - ❑ Bad fan or motor bearings
    - ❑ V-belt improperly tensioned
    - ❑ Undersized or oversized fan
    - ❑ Fan wheel out of balance
    - ❑ Air noise

5. If you identify several possible causes of a problem, start with the least expensive solution first. Check it and then go to the next least expensive.

6. Know your limitations. If you have run through a troubleshooting procedure and cannot identify the cause of the problem, check with your supervisor. If you have done your troubleshooting thoroughly, it is no reflection on your ability to say you cannot identify the cause. Some fan problems are very complex and require a specialist or engineer.

SYMPTOMS AND CAUSES

Complaints about a system are usually in one of the following categories:

- Noise and vibration
- Air noise
- Air delivery

Remember that complaints are usually about the symptoms—not the causes. Causes are usually mechanical problems and may be located in either:

- The fan
- The HVAC system

Infrequently the cause may be from the original design. Some fan types are more noisy than others. However, fan wheels and housings are designed for minimum air noise. Fan wheels and sheaves are dynamically balanced to minimize vibration. The HVAC system designer can reduce air noise by selecting the proper type and size of fan. A fan is usually most quiet when it is operating at the point on the fan curve of peak mechanical efficiency.

NOISE AND VIBRATION

Noise and vibration are usually caused by the fan, but they can be carried through the system by the air. They can also be caused by something in the duct system, such as a loose damper. In some cases, excessive air velocities in the duct can cause vibrations in the system that would not show up under normal air speeds.

Just by listening and touching you can usually determine whether the cause is in the fan or farther down in the duct system. If the cause is the fan, determine whether the noise and vibration is caused by a fan problem or whether it is simply normal fan vibration that has not been properly isolated. To do this, check the fan isolation devices.

Fan Isolation Devices

If the fan is on an **inertia base** (See Chapter 7, page 73), check that all the **isolation springs** are operating freely. The springs must not be touching any metal or concrete, and they cannot be bottoming out. If the springs are not adjusted properly, the fan can rock. If one part of the inertia base seems to be vibrating more than other parts, tighten the spring tensions until the vibration is minimized. Check the manufacturer's instructions to see if the springs are installed and tensioned properly.

For fans with **vibration dampeners** (See Chapter 7, page 71), check to make sure that the fan base is not touching any metal.

Fan vibration can also be transmitted due to improper isolation from the duct, electrical conduit, or piping. Check all these to see that they are properly isolated from the fan. Most fans have built-in flexible connections to isolate the fan vibration from the housing or from the duct connection. Check to make sure that shipping blocks under the fan base, or a shipping plate for the flexible connection have been removed.

Large fans may rock considerably during start-up and cause short-term noise or vibration by exceeding the limits of flexible connections to such things as duct and electrical conduit.

Resonance

A form of vibration is **resonance**—also called **sympathetic vibration**. If a fan vibrates, it does so at a regular frequency. Sometimes the frequency of the fan vibrations and the frequency of a part of the building (such as the fan room floor) happen to be the same (or an octave or more apart). When this occurs, the vibrations of the two components feed on each other and amplify the vibration. If vibration is excessive, resonance should be considered.

Resonance can be determined by changing the fan RPM by at least 10%. If the vibration is reduced considerably, resonance is the cause. If the vibration remains the same, then it probably is a mechanical vibration problem.

The solution to resonance is to change the frequency of one of the components. It may mean changing the fan speed or adding stiffeners to the building component. These solutions require engineering decisions.

Fan Vibration

Fan vibration problems are often from a fan mechanical problem:

- ❑ The fan may be operating at a higher RPM than it was designed for. Measure the fan RPM and check (or have your supervisor check) the RPM against the limits on a manufacturer's fan chart.
- ❑ Improper alignment is often a cause of vibration. Check the following:
    - ❑ The fan base is level.
    - ❑ The fan is aligned with the duct.
    - ❑ The fan shaft and motor shaft are parallel.
    - ❑ The motor sheave and the fan sheave are aligned.
    - ❑ The V-belts are properly aligned.
    - ❑ For direct drive fans, the motor and the fan shafts must be aligned.

- Set screws may be loose or missing.
- Bad fan bearings can cause noise and vibration on new fans as well as old ones. Bearings may be flattened if a large fan was lifted or tied down using the fan shaft or if a fan was stored for a long time without rotating the fan wheel occasionally. Sometimes you can feel the thump of a bad bearing when the fan wheel is rotated slowly. A stethoscope on the bearing when the fan is running will usually pinpoint the problem.

Motor Vibration

The fan motor may be a cause of vibration. Misalignment of the motor, bad bearings, and the motor out of balance can all contribute to vibration.

Fan Wheel Problems

Many different mechanical problems are related to the fan wheel:

- The fan wheel may be out of balance. This would be rare with a new fan, because manufacturers dynamically balance each wheel. If a new fan wheel is out of balance, first check the fan shaft to see if it was bent during shipping. On an older fan, the blades of the fan wheel can become loaded with dirt and get out of balance. Also, balancing weights, placed on the wheel when it was originally balanced, may come off.
- A fan wheel can be out of alignment in the direction parallel to the fan shaft. In extreme cases, the end of the wheel can rub on the fan housing. Many large

fans have cross rods inside the fan wheel that can correct this longitudinal misalignment.

- ❑ A fan wheel may slide on the shaft so that it is not centered in the housing. Center the wheel and tighten the set screws.
- ❑ Loose or broken tie rods in the wheel (used to true up the wheel) may cause a rattle. The wheel will wobble, causing vibration.
- ❑ The fan wheel may be out of line with the inlet cone. The opening of the fan wheel and the inlet cone should be aligned.

System Noise and Vibration

Mechanical conditions in the duct can also cause noise and vibration:

- ❑ Loose damper blades or turning vanes vibrating.
- ❑ Malfunctions in terminal units such as VAV boxes, induction units, and terminal reheat units.
- ❑ Rumbling noise caused by improper reinforcement of the duct.
- ❑ Hissing noise caused by excessive duct leakage.
- ❑ Tin-canning noise caused because gage of duct is too light.

AIR NOISE

Air noise is generated by the movement of air. It may be difficult to determine whether a noise in a conditioned space is generated by the movement of air in the fan and ducts or whether it is noise generated from fan room vibration and carried by the air. For this reason, it is good practice to check for and solve vibration problems first.

Then any noise still being carried to the conditioned space is probably from air noise.

Fan Surge

Air noise can be caused by **fan surge**. A fan that is running to the left of the recommended operating range tends to surge. (See Chapter 4, page 36.) Surging is more likely in forward-curved-blade fans than in backward-curved-blade fans.

Fan curves are available that show the surge line (Chapter 4, Fig. 7). An operating point to the left of the recommended operating range will be subject to fan surge problem.

Fan surge is caused when the fan RPM has been changed or when resistance has increased due to some duct condition such as:

- ❑ Closed dampers
- ❑ Closed fire dampers
- ❑ Dirty coils
- ❑ Loose duct lining (sound attenuation material) that restricts airflow. Duct lining should not be installed immediately after the fan where air velocities are high. It should be installed after the duct increases in size and FPM is reduced.

Other Causes of Air Noise

Air noise can also be caused by the following:

- ❑ Air velocity (FPM) too high

- ❑ Dampers too close to outlets (registers, grilles, and diffusers)
- ❑ Dampers that are severely restricted
- ❑ Undersized registers or diffusers.
- ❑ Improper duct design that causes excessive air turbulence, such as square throat elbows without turning vanes, and restrictions in the duct.
- ❑ Loose duct lining that causes excessive air turbulence

If the noise is caused by duct conditions, it can possibly be located by isolating parts of the system:

- ❑ Bypass the return air system.
- ❑ Operate each duct zone alone.

Solving air noise problems caused by improper duct design is difficult. Some methods are:

- ❑ Replacing poorly designed fittings.
- ❑ Removing a length of duct and replacing it with a section of duct with duct liner (sound attenuation)
- ❑ Installing sound attenuation boxes. However, typically these greatly increase the resistance of the system and may affect air delivery.

AIR DELIVERY

A fan problem may not be either vibration or noise, but improper delivery of air. Symptoms can be such things as drafts, stagnant air in some spaces, and failure to maintain comfort conditions in the spaces.

These problems could be caused by the condition of the fan, or may be the result of some malfunction of another system component. There may be interrelated problems involving both the fan and the duct system. Thorough investigation by knowledgeable persons is often required to identify the actual causes of the problem.

The best solution is to have TAB (testing, adjusting, and balancing) performed on the entire system. The fan and duct system were designed to deliver air to each of the conditioned spaces at the required CFM, FPM, and temperature. The TAB process will adjust the fan and the duct system to deliver air to each space as close to the design conditions as possible.

REVIEW

1. What are the three common types of fan problems?

2. What is another name for resonance?

3. What is the solution for resonance?

4. What three mechanical conditions cause fan vibration?

5. There is air noise in the ductwork. What is the first thing to check?

6. A fan is surging. Where would you expect the operating point to be on a fan curve?

11 FAN MAINTENANCE

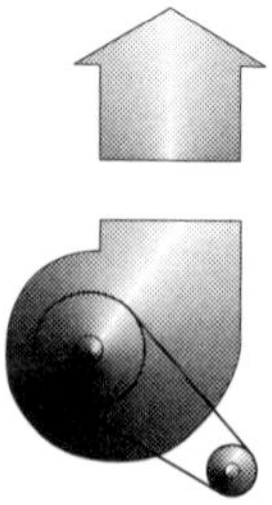

The indoor environment technician (IET) may need to provide maintenance for the fans in an HVAC system.

Fans should be inspected and serviced at least four times a year. However, the reality is that many fans are not inspected from the day they are installed to the day they break down. If you are a building engineer, you should have a preventive maintenance schedule for fans and other HVAC equipment where service dates can be entered. In addition to the regular scheduled inspections, inspect the fan every time you change filters. If you are an outside service technician, you should leave a record of your work—such as a signed and dated tag or a maintenance schedule attached to a safe spot near the fan.

SAFETY

Follow all the safety precautions given in Chapter 2. Review that chapter because it is essential for your life and safety. Especially remember:

- ❑ Lock out and tag the main switch.
- ❑ Locate the disconnect switch within sight of the fan and turn it on and off to make sure the power is off.
- ❑ Test circuits for voltage and to be sure the circuit is open.
- ❑ Avoid getting trapped in the plenum of a large HVAC system. Take precautions and let someone know where you are. Carry a cell phone.

PERSONAL RELATIONS

If you are working for an outside contractor, you need to be aware of the impact your words and actions on the job can have on other people. Your first responsibility is to your employer, but you also affect others. The positions you commonly can affect are:

- Your employer
- The building engineer
- The building owner
- The HVAC contractor who installed the system

Generally, problems are minor and you remedy them as part of your work (for example, tightening loose set screws and replacing worn V-belts). However, you can run into problems that require considerable expenses, such as replacing a motor or replacing fan bearings. Report these problems to your immediate supervisor. Your employer may have to deal with considerations such as warranties or approval for performing the work.

Use good judgment when discussing problems. If you are an outside technician, a casual remark to the building owner can create problems. What you see as a relatively small problem can be seen by the building owner as an enormous problem. Your comment may reflect on the ability of the building engineer, on the installing contractor, or on your own employer. Also, you will have made an enemy if a problem comes to the engineer as a surprise from the building owner because you have gone over the engineer's head. Also, your mention of a problem can be taken as a criticism of the work of the maintenance staff.

In general, it is good policy to discuss possible problems only with your immediate supervisor. If problems are brought to the attention of the building engineer, it should be done in a constructive and supportive way. The building maintenance staff can remedy some things.

MAINTENANCE

Maintenance should be done on a regular basis—every three months is recommended.

A good maintenance program should include an **inspection checklist**, such as that in Fig. 1, for the fan as well as for every other building component that needs regular inspection and maintenance. The checklist is a list of reminders of items that the technician should look for. It should be posted in a safe place in the mechanical room.

A **maintenance schedule** for a piece of equipment should also be posted in the mechanical room. It should include information on scheduled service from the manufacturer's information.

A **logbook** should be maintained to serve as a history of the equipment. Every inspection and maintenance should be recorded in the logbook. The logbook should have a space for notes on repairs done and items to watch for. It should also indicate the expected life of the equipment so that replacement can be anticipated. A copy of the inspection checklist should be included in the logbook.

In addition to regular maintenance recorded in the logbook, a good technician should inspect equipment daily as it is encountered. Use your senses of sound, sight, and smell to discover any possible problems developing. Catching a problem early can prevent more serious problems later.

Scheduled servicing of a fan is a relatively simple process. It consists of **inspection** and **lubrication**. But simplicity does not mean it is unimportant. If you do not understand what you are doing, it can unnecessarily cost the building owner

SAMPLE INSPECTION CHECKLIST

Plenum
- Clean and clear of debris

Fan Wheel
- Dirt build-up on fan wheel
- Balance weights on blades loose or missing
- Excessive wear or corrosion apparent on blades
- Bolts and set screws loose or missing

V-Belt Drives
- Signs of wear on sheaves
- Signs of wear on belts
- V-belt alignment and tension

Noises
- Unusual noise, such as rattles, knocks, rubbing, or grinding sounds
- Excessive air noise
- Fluctuating air noise

Other
- Feel for excessive vibrations
- Fan and motor bearings running smoothly—no noise or grinding sounds
- Condition of scroll (broken welds, etc.)

Lubrication
- Lubricate fan bearings
- Lubricate motor bearings

Fig. 1: Inspection checklist for a fan

thousands of dollars, and it can damage your reputation as a skilled and reliable technician.

Inspection

Inspect the fan for any problems or potential problems. The checklist (Fig. 1) indicates typical items to be checked. If you detect any problems or potential problems, you should remedy the simple ones. If the solution requires considerable expense, you need to report it.

LUBRICATION

Proper lubrication is essential for trouble-free fan operation. Inadequate or improper lubrication is the greatest cause of fan breakdown. Manufacturers provide lubrication instructions. Large fans generally have a decal with lubricating instructions. However, you must establish the proper time interval between lubrications according to the particular fan operating conditions, as explained in the next section.

Grease Bearings

Almost all large fans have **grease bearings** with grease fittings. Fan bearings are filled with grease at the factory. At fan start-up, grease may be pushed from the bearing seals for a little while. This is normal. Wipe off the excess, but do not add grease, because the bearing is working off excess grease.

The manufacturer's recommended time interval between lubrications should be considered as a suggested minimum. You need to establish the proper time interval for the job conditions because it is affected by such things as:

- Shaft speed
- Number of daily hours of operation
- Surrounding air conditions, such as temperature, humidity, or corrosive or abrasive materials in the air.

Vertical shaft bearings require lubrication about twice as frequently as horizontal shafts.

LUBRICATION FREQUENCY IN MONTHS					
	Operating RPM				
Shaft size in inches	**500**	**1000**	**1500**	**2000**	**2500**
5/8 - 1	6	6	6	6	4
1 1/8 – 1 1/2	6	6	6	4	4
1 5/8 – 1 15/16	6	6	6	4	4
2 – 2 1/2	6	6	4	4	2
2 11/16 – 3 3/16	6	4	2	2	1
3 7/16 – 3 13/16	6	4	2	1	Consult manufacturer

Fig. 2: General recommended lubrication schedule for horizontal shafts based on RPM and shaft size

Figure 2 is a general suggested bearing lubrication schedule for horizontal shafts according to fan shaft size and RPM. You should establish the proper interval for the installation by first following the recommended interval from the manufacturer or from this general schedule.

Be sure to follow the fan manufacturer's instructions regarding the type of grease to use. Greases have different bases, such as synthetic base, lithium base, and sodium base. Mixing these is asking for trouble. Different greases may be incompatible and could result in rapid deterioration or breakdown of the grease. If the manufacturer's instructions are not available, use a good quality lithium-based grease with the following specifications:

- NLGI Grade 2 with rust inhibitor and antioxidant additives
- Minimal viscosity of 500 SUS at 100°F

Use a grease gun to pump grease into the grease fitting:

- Use only a moderate pressure so that you do not break the grease seals.

- ❑ If the bearing has a drain plug, open it and pump grease until new grease comes out the drain.
- ❑ If there is no drain plug, pump grease until new grease comes out the seals.
- ❑ Examine the old grease that is pumped out of the bearing:
    - ❑ If it is only slightly darker than the new grease, the time interval is about right.
    - ❑ If it is very much darker than the new grease, more frequent lubrication is needed.
    - ❑ If it has metal particles, there are bearing problems.

A properly lubricated bearing should be only about $\frac{1}{3}$ to $\frac{1}{2}$ full of grease. To assure this, after the bearing is filled with grease, open the drain plug (if there is one) and run the fan for about five minutes. Then replace the plug. If there is no drain plug, run the fan until no grease comes out of the grease seals.

Be sure to wipe off any grease from the outside of the bearings. However, leave a little grease at the top the grease fitting. This prevents any dirt or moisture from entering the bearing.

Oil Bearings

Oil bearings are seldom found on large HVAC fans, but they may be used on smaller fans. To add oil:

- ❑ Have the fan stopped and add the oil slowly.
- ❑ Fill the oil cup to just below the top.
- ❑ If there is an oil gage, add the oil into the gage slowly until it reaches the proper level.
- ❑ Wipe off all excess oil from drain plugs and fillers.
- ❑ After the fan has run for a few minutes, stop it and check the oil level.

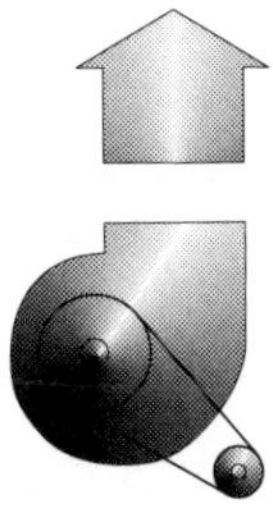

REVIEW ANSWERS

Chapter 1

1. E
2. H
3. B
4. G
5. C
6. I
7. A
8. F
9. D
10. J
11. L
12. M
13. K

Chapter 2

1. Never, under any conditions
2. Place a barrier (such as a ladder) in front of the V-belt drive as a warning.
3. Lock out/ Tag out
4. The belt can move and crush your fingers between the belt and sheave.
5. Let someone know you are in the plenum.
   Carry a flashlight.
   No loose, long hair.
   Be aware of hazards to the head.
   Carry a cordless phone.

Chapter 3

1. B
2. H
3. E
4. E
5. C
6. G
7. D
8. F
9. I
10. A
11. Single width, single inlet
12. Double width, double inlet
13. Bottom Horizontal – CW BH
14. Up Blast – CW UB
15. Down Blast – CW DB
16. Top Angular Down – CW TAD
17. Bottom Angular Up – CW BAU
18. Top Angular Up – CW TAU
19. Top Horizontal – CW TH

Chapter 4

1. 5.46 Bhp
2. 2291 RPM
3. 5361 CFM
4. Class II
5. 8.67 Bhp
6. 3400 FPM
7. 2676 RPM
8. Class II

Chapter 5

1. System effect
2. Friction loss
   Dynamic loss
   Pressure loss

3. Friction loss
4. Dynamic loss
5. Pressure loss
6. Turning vanes
7. Static regain
8. 30"
9. 42"
10. 19.7"
11. Evase′

Chapter 6

1. A
2. I
3. D
4. G
5. B
6. J
7. C
8. F
9. E
10. H
11. From the manufacturer's catalog
12. They can be made to any length needed.
13. 4 (rounded from calculated 3.214)

Chapter 7

1. E
2. G
3. H
4. F
5. A
6. D
7. B
8. C

Chapter 8

1. A. 627 RPM
   B. 5.9" (Nearest sheave is 5.8")
2. A. 734 RPM
   B. 4.8"
3. 10,821.8 rounded to 10,820 CFM
4. 7101 rounded to 7100 CFM
5. 17.8" (Nearest sheave is 18.4". This gives an RPM that is within 97% of the required RPM.)

Chapter 9

1. A. 14,175 CFM
   B. 1.14"wg
   C. 4.52 Bhp
2. 496 RPM
3. A. 650 RPM
   B. 14,996 rounded to 14,995 CFM

Chapter 10

1. Vibration
   Air noise
   Air delivery
2. Sympathetic vibration
3. Change the frequency of one of the components.
4. Fan operating at RPM that is too high
   Improper alignment of the fan
   Bad fan bearings
5. Check the fan for vibration and noise.
6. To the left of the recommended operating range.

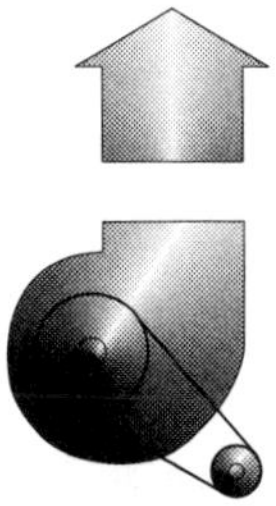

GLOSSARY

AMCA—Air Movement and Control Association

AMD (air moving device)—Fan

Atmospheric pressure—The pressure of the air around us (14.7 psi at sea level)

Axial fan—Propeller type fan with airflow parallel to the fan shaft

Belt guard—A safety screen over the V-belt drive to protect people and property from the moving parts

Blower—Centrifugal fan

Brake horespower (Bhp)—A measure of the load imposed on a motor by a fan or other device

Bushing—Metal fitting used to adapt the hole size of the sheave to the shaft diameter

Centrifugal fan—Fan that moves air by the centrifugal force of a fan wheel

CFM (cubic feet per minute)—A measure of the quantity of air moving in a duct

Cut-off—Metal plate that prevents the air discharged from the fan wheel from being drawn back into the scroll

Datum diameter—The working (effective) diameter of A, B, C, and D sheaves

Drive sheave—The sheave on the motor shaft

Driven sheave—The sheave on the fan shaft

Duct liner—Insulation on the inside of the duct to reduce heat loss or heat gain and to deaden sound

Equipment room—A room where the central HVAC equipment (piping, pumps, fan plenum, etc.) is installed

Evasé—A specially designed expanding duct transition connecting a fan outlet to a supply duct

Fan curve—Graph that shows the performance characteristics of a type, model, or size of fan

Fan efficiency—Relationship between the useful work produced by the fan and the total work input by the fan motor, expressed as a percentage:

$$\% \text{ Fan efficiency} = \frac{\text{Useful work output}}{\text{Total work input}}$$

Fan laws—Equations that relate the performance variables (CFM, RPM, SP, Bhp) of a fan in a fixed system—used to compare performance before and after a change is made

Fan plenum—A large sheet metal enclosure for the fan and other items in the equipment room

Fan surge—Fluctuation of CFM and SP in a fan and ductwork resulting in surges of air. Caused by operating outside the recommended operating range of the fan

Fan table—Table that shows the performance characteristics of specific models and sizes of fans

FPM (feet per minute)—A measure of the velocity of air in duct

Horsepower (**Hp**)—A measure of the work capacity of a motor

Housing—Fan casing

Inches water gage (+)—A measure used for measuring air pressure in duct

Inertia base, inertia block—Concrete slab attached to the floor by heavy coil springs and used as the base for a large fan and motor

Inertia—Tendency of a object at rest to remain at rest or of an object in motion to remain in motion

Inlet cone—A cone-shaped device designed to draw air into the fan efficiently

Inlet vanes—Vanes at a fan inlet that can be adjusted to control the volume of air that a fan produces

Isolation blocks—See **Vibration dampeners**

Key and keyway—A keyway is a groove in the shaft and in the hub of a sheave or its bushing, and the key is a rectangular piece of steel that fits into the keyway to provide a positive connection

Manometer—Instrument that indicates air pressure in a duct

Outlet dampers—Adjustable blades on the fan discharge used to control the volume of air that a fan produces

OV (**outlet velocity**)—Velocity of the air at the fan outlet

Pitch—Angle of an object, such as the angle of the vanes for a vaneaxial fan

Pitch diameter—The working (effective) diameter of the sheave

Pitch length—Length of the belt at the point where the belt contacts the pitch diameter of the sheave

Plug fan—Centrifugal fan with no housing, designed to be used in a fan plenum

psi (pounds per square inch)—A measure of pressure (usually water or steam pressure)

Resonance—Vibration of the fan and some part of the building at the same frequency (or an octave or more apart)

RPM (revolutions per minute)—A measure of rotational speed

Scroll—Fan casing

Seismic restraints—Devices to prevent a fan from tipping or moving off its mountings during an earthquake

Sheave—Pulley on the motor shaft or fan shaft with a V-groove to match the V-belt

Shroud—A plate that surrounds an axial fan to guide the air through the fan blades

Slide rails—Motor mounts that allow a motor to slide back and forth to adjust the tension of the belts

Sound attenuation—Any method designed to lessen the noise made by a fan

Sound attenuator—Manufactured device installed in duct in order to deaden noise

Static pressure (SP)—Pressure of air in duct that is effective in all directions (like the air that keeps a balloon expanded)

Static regain—Process of converting the velocity pressure of air in a duct to static pressure

Sympathetic vibration—See **Resonance**

System effect—Negative effect on fan performance when a fan is attached to a duct system

System resistance—Amount of resistance to airflow imposed by the duct system

Tip speed—The speed of a point on the outer edge of a fan wheel

Torque—A force transferred by a twisting or rotating motion

Total pressure (**TP**)—The total of static pressure and velocity pressure in a duct

Tubeaxial fan—Axial fan with a tube-shaped housing

Tubular centrifugal fan—Centrifugal fan with a tubular housing instead of a scroll

Turning vanes—Curved blades installed in a duct elbow to turn the direction of airflow more efficiently

Vaneaxial fan—Tubeaxial fan with guide vanes to straighten the airflow and improve fan efficiency

Variable frequency drive (**VFD**)—Electronic device in a cabinet used to vary the speed of the fan by varying the frequency and voltage of the electrical power to the motor

V-belt—The belt that fits on the sheaves and transmits power from the motor to the fan

V-belt drive—The system that transmits power from the motor to the fan

Velocity pressure (**VP**)—Pressure of air in duct that results from the air being in motion

Vibration dampeners—Springs or rubber blocks that keep vibration of the fan from being transmitted to the building

Vortex—See **Inlet cone**

Wheel (of a fan)—A framework with blades and backplate that moves air out of the fan

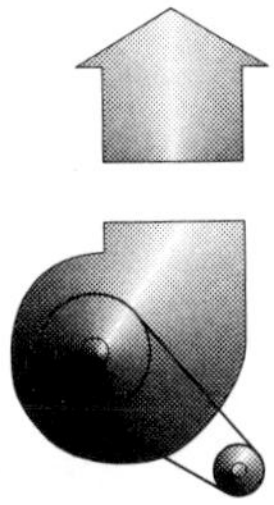

INDEX